FOREWORD

S0-ADA-903

This Repair Master contains information and service procedures to assist the service technician in correcting conditions that are not always obvious.

A thorough knowledge of the functional operation of the many component parts used on appliances is important to the serviceman, if he is to make a proper diagnosis when a malfunction of any part occurs.

We have used many representative illustrations, diagrams and photographs to portray more clearly these various components for a better over-all understanding of their use and operation.

IMPORTANT SAFETY NOTICE

You should be aware that all major appliances are complex electromechanical devices. Master Publication's REPAIR MASTER® Service Publications are intended for use by individuals possessing adequate backgrounds of electronic, electrical and mechanical experience. Any attempt to repair a major appliance may result in personal injury and property damage. Master Publications cannot be responsible for the interpretation of its service publications, nor can it assume any libility in connection with their use.

SAFE SERVICING PRACTICES

To preclude the possibility of resultant personal injury in the form of electrical shock, cuts, abrasions or burns, etc., that can occur spontaneously to the individual while attempting to repair or service the appliance; or may occur at a later time to any individual in the household who may come in contact with the appliance, Safe Servicing Practices must be observed. Also property damage, resulting from fire, flood, etc., can occur immediately or at a later time as a result of attempting to repair or service — unless safe service practices are observed.

The following are examples, but without limitation, of such safe practices:

1. Before servicing, always disconnect the source of electrical power to the appliance by removing the product's electrical plug from the wall receptacle, or by removing the fuse or tripping the circuit breaker to OFF in the branch circuit servicing the product.

NOTE: If a specific diagnostic check requires electrical power to be applied such as for a voltage or amperage measurements, reconnect electrical power only for time required for specific check, and disconnect power immediately thereafter. During any such check, ensure no other conductive parts, panels or yourself come into contact with any exposed current carrying metal parts.

2. Never bypass or interfere with the proper operation of any feature, part, or device engineered into the appliance.

3. If a replacement part is required, use the specified manufacturers part, or an equivalent which will provide comparable performance.

4. Before reconnecting the electrical power service to the appliance — be sure that:

    a. All electrical connections within the appliance are correctly and securely connected.
    b. All electrical harness leads are properly dressed and secured away from sharp edges, high-temperature components such as resistors, heaters, etc., and moving parts.
    c. Any uninsulated current-carrying metal parts are secured and spaced adequately from all non-current carrying metal parts.
    d. All electrical ground, both external and internal to the product are correctly and securely connected.
    e. All water connections are properly tightened.
    f. All panels and covers are properly and securely reassembled.

5. Do not attempt an appliance repair if you have any doubts as to your ability to complete it in a safe and satisfactory manner.

MASTER PUBLICATIONS

HUNTINGTON CITY-TOWNSHIP
PUBLIC LIBRARY
200 West Market Street
Huntington, IN 46750

1

TABLE OF CONTENTS

We gratefully acknowledge the cooperation extended by the many manufacturers to Master Publications, in the preparation of this repair-master.

SERVICE CHECK LIST

The following diagnosis charts are intended to be only a starting point in repairing ice makers. The diagnosis charts can deal only in generalities. To effectively service any ice maker the serviceman must thoroughly understand the mechanical functions and the electrical circuitry of both the refrigeration system and the ice maker.

A considerable amount of time can be saved if a serviceman will allow himself the time to properly analyze the probable cause of malfunction, before proceeding to remove parts. Proper installation of both the ice maker and the refrigerator is very important. Be sure that cold water is being supplied unrestricted to the ice maker and the operation of the ice maker in relation to the refrigerator is understood.

Always visually check for burnt or broken terminals and connections before using test equipment. The power source should be disconnected before attempting to remove any electrical part. If a voltmeter or test lamp is being used for testing, the power must be reconnected and extreme caution must be observed.

The ice makers portrayed within this Repair-Master and the accompanying text covers the following manufacturers of refrigerator equipment. If the type of ice maker used is unknown to the service man, it would be well to select points of recognition on the various illustrations so a determination may be made as to what text to follow.

ADMIRAL	GIBSON
AMANA	HOTPOINT
COLDSPOT	KELVINATOR
FRANKLIN BRANDS	NORGE-FEDDERS
Marquette	PHILCO-FORD
Wizard	PENNCREST
Bradford	SIGNATURE (Mont. Ward)
Coronado	WESTINGHOUSE
Etc.	WHIRLPOOL
FRIGIDAIRE	Many others not listed.
GENERAL ELECTRIC	

WHIRLPOOL DESIGN

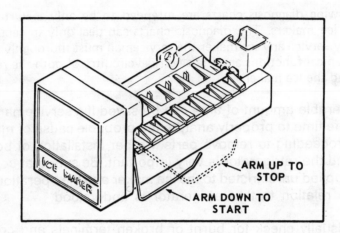

EARLY SERIES

WHIRLPOOL TYPE

SERVICE ANALYSIS CHART

The following service analysis chart is prepared in a manner to simplify the location of problems that may occur in the ice maker by the **symptom**, the **possible causes**, and the **tests and corrections** required.

SYMPTOM: Will Not Start

POSSIBLE CAUSE	TESTS AND CORRECTIONS
No electrical power.	Check for power at door switch with plunger depressed and at ice maker leads. Replace switch if defective or correct cabinet wiring.
Shut-off arm not properly positioned.	Check if arm is in lowest position and that rubber stop is lowered.
High freezer temperature.	Check mold temperature at a mounting screw. If it is above 15° F., check analysis chart in Refrigerator Manual.
Inoperative thermostat.	If mold temperature is below 15° F., place a jumper between the common and cold terminals. If ice maker starts, replace thermostat.
Open limit switch.	Check for continuity. Replace if open.
Shut-off arm switch not properly positioned or open.	Check position of switch and adjust if necessary. Check for continuity when arm is in lowest position and replace if open.
Inoperative water solenoid switch.	With ejector blades in starting position, check terminals "C" and "NC" for continuity. Replace switch if open.
Inoperative motor.	Direct test motor. If it fails to start, replace motor.

SYMPTOM: Does Not Complete Cycle

POSSIBLE CAUSE	TESTS AND CORRECTIONS
Holding switch not properly positioned or open.	If ejector blades stop at about the 12 o'clock position, check terminals "C" and "NO" for continuity. Adjust or replace switch if open.
Inoperative mold heater.	If ejector blades stop at about the 4 o'clock position, check heater for continuity. Replace mold and heater assembly if heater is open.
Inoperative reset heater.	If ejector blades stop at about the 7 o'clock position, check continuity of reset heater. If open, replace heater and conductor assembly. Make sure heater is properly wrapped around conductor.
Inoperative thermostat.	If ejector blades stop at about the 7 o'clock position and reset heater is good, check for continuity between the common and warm terminals. Replace thermostat if open.
Open limit switch.	Check for continuity. Replace if open.

SYMPTOM: Does Not Stop At End Of Cycle

POSSIBLE CAUSE	TESTS AND CORRECTIONS
Holding switch improperly position ed.	Check adjustment with relationship to timing cam and adjust if necessary.
	With switch properly adjusted and ejector blades in starting position, check "C" and "NO" terminals for continuity. Replace switch if closed.

SYMPTOM: Continues To Eject With Full Ice Bin

POSSIBLE CAUSE	TEST AND CORRECTIONS
Shut-off arm or switch improperly positioned.	Check that shut-off arm is linked properly so when arm is raised, switch will open. Adjust switch if required.
Shut-off switch stays closed.	Check continuity with arm raised and switch adjusted. If closed, replace switch.

SYMPTOM: Undersize Ice Pieces

POSSIBLE CAUSE	TESTS AND CORRECTIONS
Mold not level.	Check level and adjust as required.
Ejector blades too high.	Check that blades stop at proper height and that ice does not fall into mold. Adjust blade height at timing cam if required.
Inadequate water supply.	Check that supply line and water valve strainer are completely open and that adequate pressure is maintained. Clear restrictions or advise customers accordingly.
Improperly positioned water solenoid switch.	Test cycle ice maker and measure water fill. Adjust switch as required.
Short cycling thermostat (indicated by ice shells or hollow ice pieces).	Check that some part of the thermal slug in mold support is in contact with water in mold cavity. If all of the slug is above the water, the mold is not level, the water supply is inadequate or the water solenoid switch is not properly adjusted. Check thermostat bond to heat conductor and heat conductor bond to thermal slug in mold support. Assure good thermal contact both places. Check thermostat calibration by replacing with new part.

SYMPTOM: Water Spills From Mold

POSSIBLE CAUSE	TESTS AND CORRECTIONS
Poor mold gasket seal.	Check gasket between mold and support. Replace if leaking. (A very small leak may be corrected by a seal of silicone grease.)
Mold not level.	Check level and adjust as required.
Water inlet tube not properly positioned.	Check that inlet tube and fill trough fit properly and water does not leak during fill cycle. Adjust fit if required.
Water valve leaking.	Check that water does not enter mold after cycle is completed. Replace valve if leaking, check if water pressure is proper.
Improperly positioned water solenoid switch.	Test cycle ice maker and check that water fill does not exceed the required volume. Adjust switch if required.
Ice maker does not stop at end of cycle.	Refer to similarly described symptom listed elsewhere in this chart.
Short cycling thermostat.	Refer to similarly described cause under "Undersized Ice Pieces Symptom" listed elsewhere in this chart.
Ice stored on ejector blades contacting mold and melting.	Check for proper position of ejector blades and adjust as required.

SYMPTOM: Water Fails To Enter Mold

POSSIBLE CAUSE	TESTS AND CORRECTIONS
Blocked water supply.	Check that water line, valve, and valve strainer are open. Remove restrictions, open valve, or instruct customer accordingly.
Slow leak in water valve.	Observe inlet tube and fill trough for ice. If obstructed with ice, check water valve for slow leak. If valve leaks, check for proper water line pressure. Replace water valve if pressure is within specifications.
Inoperative solenoid coil.	Check terminals for continuity. Replace coil if open or shorted.

SERVICE PROCEDURE

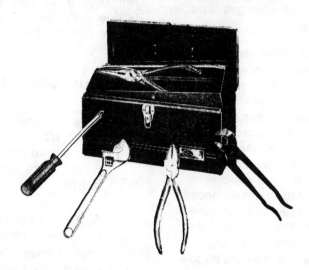

COMPONENT DESCRIPTION

Before attempting to service an automatic ice maker of any make, the serviceman should be equipped with the proper tools. Many of these are special tools designed to do a particular job quickly and to protect various parts from damage. Special tools used to service all makes include a test lamp or voltmeter, a continuity tester or ohmmeter, and a wattmeter. Proper use of these special tools will help make fast efficient diagnosis and service much easier.

As a safety precaution, ALWAYS disconnect electrical power from the automatic ice maker before attempting to remove any parts. For testing purposes, the power cord can again be plugged into a live receptacle after the necessary parts are removed.

It is advisable to make certain that the water supply faucet or valves are closed and the ice mold is drained of all water, if any parts are to be removed or disconnected in the water system.

Due to the large number of types covered in this manual, no attempt will be made in this section to give a complete detailed step-by-step procedure on disassembly of each individual model. Instead we will give the service procedure and functional description of the various components as used on most models. In a few cases these components may not be identical to the ice maker being serviced but their function, as well as service procedure, will be the same.

THE AUTOMATIC ICE MAKER—THE WHIRLPOOL TYPE

This Repair-Master covers the operation, diagnosis, and service procedures pertaining to the factory installed and add on ice maker installed by the dealer or service technician. This style ice maker has been in use many years, and with some changes and modifications is still with us today.

To effectively service this or any ice maker, the technician must first determine if the problem is within the ice maker unit, or due to malfunction of the refrigeration system itself. Servicing of the ice maker and the related circuits will be explained in this Repair-Master. A comprehensive look at servicing the refrigeration system may be found in the Master Publication Refrigeration Repair-Master #7551 & 7552 available at your refrigeration parts supplier. Servicing the ice maker will be relatively easy, once you understand its basic operation and become familiar with the few components that on occasion may have to be adjusted or replaced. Whether you are now repairing an ice maker or not, it would be to your advantage to study this Repair-Master so you may easily handle all service calls, be it for the refrigerator itself, or the ice maker.

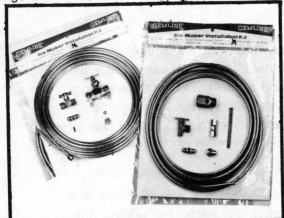

INSTALLATION
Figure W-1

Proper installation of an automatic ice maker should begin with the selection of the necessary valve, filter, fittings and length of tubing. A common service practice is to use a ice maker installation kit such as a Gemline IMT-25CFSV or IMU-25C (Figure W1) depending upon where the most convenient and practical source of water supply exists. The IMT series kit offers a saddle clamp and shut off valve easily adapted to a ³⁄₈" to 1³⁄₈" water line, while the IMU series offers an adapter tee and shut off valve for undercounter installation. Both kits include 25' length of copper tubing. A water filter of the charcoal cartridge type is normally placed in the line ahead of the water inlet valve. The Gemline IWF 101 filter is complete

with charcoal cartridge No. IWC 201. A filter of this type is highly recommended to assure clear, oderless, and tasteless ice cubes. The filter cartridge can be easily changed when clogged or no longer effective.

Always allow enough tubing, approximately eight to ten feet, that can be coiled and left behind refrigerator. This will afford ample length to move refrigerator out away from wall for future service or cleaning.

WATER LINE CONNECTIONS

a. Move the rubber stop up to the "off" position, see Figure W2. This will keep the refrigerator water valve closed and prevent the operation of the ice maker. Then turn off the water supply at the source

b. Determine the water source in relation to the refrigerator, connect the tubing through a water shut off valve. The valve should be accessible in case of an emergency. If in the kitchen, a proper water source would be from under the cabinet, and using the adapter ahead of the bib valve that shuts the cold water off.

c. Route the water line so it can be positioned behind the refrigerator, and hidden behind the cabinets or under the counter.

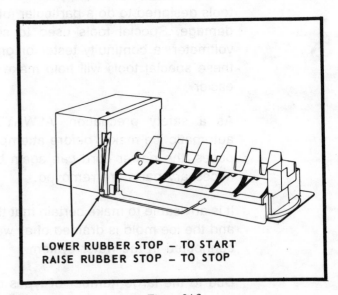

LOWER RUBBER STOP – TO START
RAISE RUBBER STOP – TO STOP

Figure W-2

d. Shape the remaining tubing behind the refrigerator into a large coil of two or three turns. Clamp the tubing to the baseboard or wherever convenient, so when it becomes necessary to move the referigerator away from the wall for cleaning purposes, it will not be a problem. (Figure W3).

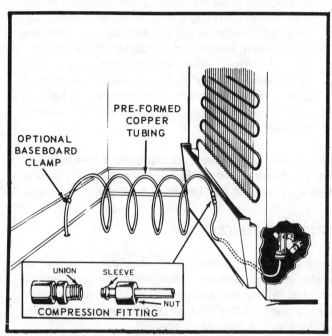

Figure W-3
WATER SUPPLY CONNECTIONS

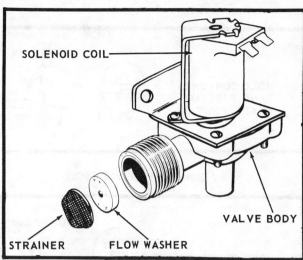

Figure W-4
WATER VALVE AND SOLENOID ASSY.

COMPONENT DESCRIPTION AND CHARACTERISTICS

The water connections having been made, the refrigerator plugged in and turned on, and the ice maker in operation, the customer should be advised to discard the first and second trays of ice because of foreign matter that may have found its way into the water.

A-1 In the "freeze cycle" water fills the ice mold. The shut off arm is in the lower position, the contacts of the shut off switch are closed. If ice has been made previously, it will lie on top of the "ejector blades" (see figure A-1).

A-2 When the ice in the mold has frozen, and the temperature has been lowered to about 16° F, the contacts of a thermostat in the ice maker allow the "mold heater" to operate, along with the ice maker motor. The motor rotates the "ejector blades" and a timing cam. As the ejector blades rotate upwards, the ice on the blades is lifted and they fall into the ice bin (Figure A-2).

One section of the timing cam, during the same period of rotation, closes the contacts of the holding switch. Still another section of the timing cam causes the shut off arm to open the contacts of the shut off switch (Figure A-2). The ice maker will continue to operate when the shut off switch is open, because the holding switch has been closed. The ice maker is now controlled by the holding switch.

A-3 The ejector blades continue in their rotation until they contact the ice in the molds (Figure A-3). The motor is stalled until the mold heater releases the ice from the mold. A stall type motor is used, and if it is stalled for a few seconds or a few minutes, both are a normal function and the motor will not be hurt.

e. Before making a connection to the refrigerator, turn on the water and flush the line out. Connect tubing to the refrigerator.

f. Turn water on and check for leaks. Wait a few moments as there may be air in the line and a leak will not show up for a few minutes.

g. Move refrigerator into position, guiding the line and tubing coil so it will not interfere with the condenser.

h. Instruct the user and point out location of the shut off valve.

i. Turn the ice maker to the "On" position. It will be in a condition to make ice when the freezer door is closed. If there is a reset button on the ice maker, follow the instructions.

j. The ice maker operation will go off every time the freezer door is opened. This is controlled by the push switch on the freezer side, and is a protective device.

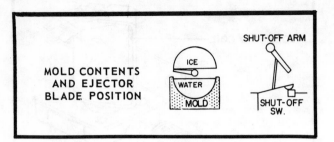

Figure A-1 _ FREEZING CYCLE

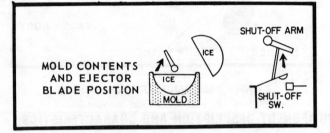

Figure A-2 EJECTING ICE INTO BIN

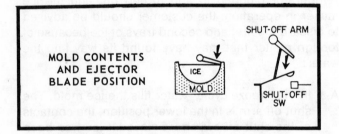

Figure A-3 EJECTOR BLADES CONTACTING ICE IN MOLD

Figure A-4 ICE REMOVAL AND WATER REFILL

A-4 When the ice has melted free of the mold, the ejector blades continue their rotation and sweep the ice crescents out of the mold (Figure A-4). While this is happening, another section of the timing cam closes the water valve switch, energizing the solenoid and allowing water to refill the mold. The timing cam continues in its path to open the water switch contacts, thus closing off the water supply. The fill time is about 7 or 8 seconds.

Another action that takes place during this phase of the operation is that the shut off arm lowers to its original position, closing the contacts of the shut off switch (Figure A-4). Once an ejection cycle is completed, it will not start again, unless the shut off arm is lowered enough to close the contacts of the shut off switch. When the ice bin fills to capacity, the ice does not allow the shut off arm to lower. This will stop the next ice harvest until some of the ice is removed to allow the arm to lower.

The harvest cycle takes approximately 2 to 4 minutes. The ice making period is approximately 1¾ hours. The ice bin should be filled near capacity in 24 hours.

PROTECTIVE DEVICES

There are many protective devices used on the ice maker, not the least of these being the push switch mounted on the breaker strip of the freezer compartment. Power to the freezer is interrupted whenever the door to the freezer is opened.

Failure of the holding switch to open would cause a high temperature limit switch to open, interrupting the power and shutting off the mold heater.

SOLENOID COIL AND WATER VALVE

To control the flow of water, a solenoid activated water valve is used. It is mounted to the right, or sometimes the left, and just ahead of the rear legs in the machine compartment. There are some refrigerators that allow a special place in a recess to the rear of the refrigerator and away from the machine compartment. Regardless of the position mounted, they all work alike (Figure W-4).

When the ice maker calls for water, the solenoid coil, which produces a magnetic field when energized, lifts the armature from the orifice of the valve. This allows the water to flow under and through the bleeder holes in the orifice diaphragm, thus equalizing the pressure on the diaphragm and allowing the water to flow into the ice mold.

To check the valve proceed as follows;

WATER VALVE TEST

a. Disconnect existing wires to valve.

b. With a live test cord, touch both terminals of the solenoid.

c. If water flows into the ice maker, the problem is elsewhere.

d. If water fails to flow, check solenoid with a continuity tester.

e. If continuity exists, the trouble would be a stuck closed valve.

f. Remove valve, clean or replace, see section CLEANING THE WATER VALVE.

NOTE: If the condition exists that water comes in intermittently or at the wrong time, the indications are that the valve is sticking open or closed due to mechanical failure, and should be replaced or cleaned.

FLOW WASHER (Figure W5)

If the proper quantity of water is not being metered into the ice maker, water will overflow the ice mold, or you may be starving the ice mold and produce shells of ice rather than crescents or cubes. If the flow washer becomes deteriorated, such a condition would prevail. To replace the flow washer, remove the water inlet tube and brass fitting, and remove the flow washer. When a new flow washer is installed, be sure that the identifying numbers are facing the incoming water.

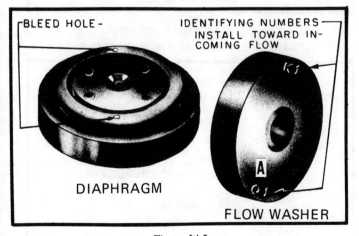

BLEED HOLE — IDENTIFYING NUMBERS INSTALL TOWARD INCOMING FLOW

DIAPHRAGM

FLOW WASHER

Figure W-5

OPERATION OF THE WATER VALVE

CLEANING THE WATER VALVE

NOTE: A good cleaning agent to use on the internal parts of the valve is a 50% solution of soldering acid (can be purchased at a local hardware store) and water. Mix solution in a glass bowl.

a. Remove valve from valve mounting bracket.

b. Remove wiring from solenoid coil.

c. Remove fill hose.

d. Remove coil from valve. Metal encased coil is a three screw mount, the encapsulated type is one screw mount, but the other two screws must be removed to open the valve. The old style coil has just a retainer at the top, note which side of this latter coil is outside or away from the valve body. It must be reinstalled in the same position as removed.

Remove the three screws or the four screws according to type of valve.

e. Remove brass cylinder and diaphragm.

f. Inspect valve seat for any nicks or worn spots.

g. Remove spring and plunger from cylinder.

h. A valve kit can be purchased. The kit contains a cylinder, plunger, spring and a diaphragm.

i. If a new kit is not readily available, dip the above parts in the solution as outlined above in NOTE. *When parts are clean of lime deposits, remove and wash with clean water.*

j. Reassemble to valve. Replace solenoid coil and install on machine.

NOTE: The diaphragm has an orifice hole and a bleeder hole. (Figure W5).
CAUTION: Avoid getting acid on your hands.

DIAPHRAGM (Water Valve)

If the water flow into the tub cannot be stopped even though the inlet solenoid is deenergized, there is a possibility that the tiny bleeder holes in the diaphragm have become clogged. See *Figure W5*. Remove inlet valve as described under *"Water Inlet Valve Text"*, thus exposing the diaphragm. Diaphragm can be cleaned in acid solution as previously outlined. By slightly stretching the diaphragm, any lime deposits will fall out of the bleeder holes. Check valve seat for nicks or rough spots. The nicks can be lapped out using a fine grade of emery cloth on a flat surface.

STOP, SHUT-OFF ARM

A rubber stop located on the vertical portion of the shut-off arm, (Figure W6),activates the shut-off switch. When the stop is in the lower position as shown, the contacts of the shut-off switch are closed, and the ice maker is in operation.

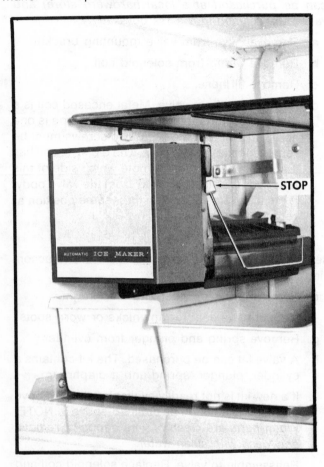

Figure W-6 – STOP IN "OFF" POSITION

ICE MOLD AND HEATER

Ice is formed in an aluminum die cast mold having crescent shaped compartments, (Figures W7-W8). Water enters at the rear, and each compartment is notched, which allows the water to flow from one compartment to another.

The heater is aluminum sheathed and is cemented in place and recessed in a groove provided,(Figure W7). When it is energized,the heat melts the outer surface of the ice and frees it from the mold, thus allowing the harvest. The ice mold and heater are one piece and must be replaced as a unit.

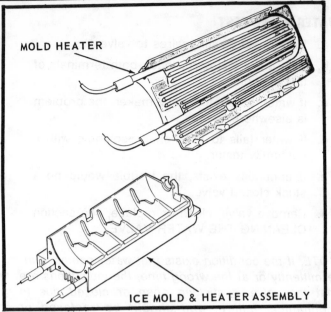

Figure W-7
ICE MOLD AND HEATER ASSEMBLY

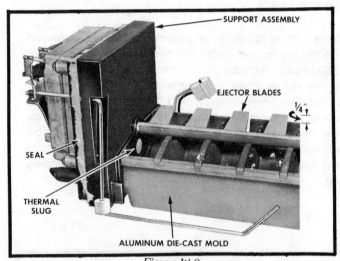

Figure W-8
SUPPORT ASSEMBLY, ICE MOLD,
AND THERMAL SLUG EJECTOR BLADES

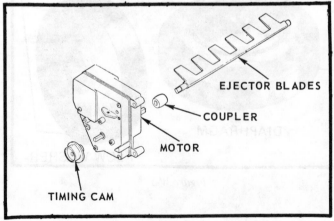

Figure W-9
EJECTOR BLADES, COUPLER, MOTOR AND CAM

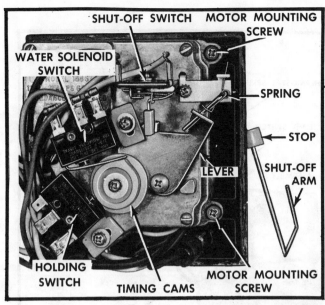

Figure W-10

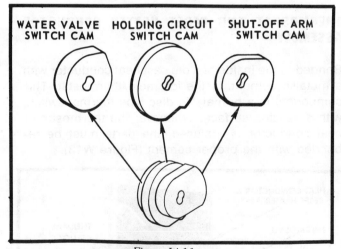

Figure W-11
THREE-IN-ONE TIMER CAM

EJECTOR BLADES (Figure W9)

The ejector blades are cut to equal size from an aluminum extrusion. The blades are designed to sweep the ice from the mold when the heater loosens it from the ice mold. The central shaft or fulcrum rides in nylon bushings and is coupled to the motor output shaft. At the completion of a cycle, the ejector blades should come to rest even with, or approximately 1/4 inch above, the left side of the mold.

If the blades of the injector stopped lower, the water would freeze to them, if they were higher the ice crescents of the last harvest would fall into the mold cavity, causing water that enters to overflow the mold and be deposited in the freezer liner.

SUPPORT ASSEMBLY AND COMPONENTS

The support assembly, made of bakelite, is attached to one end of the mold (Figure W8).

An aluminum thermal sensing device is part of the assembly. Temperature is transferred to the front ice piece, then to a heat conductor, (Figure W8), and in turn, contacts the sensing surface of a thermostat. When temperature is reduced 16° F., it is sensed by the thermal slug, which causes the thermostat contacts to reset to the cold position and a harvest cycle begins.

TIMING CAMS

Three separate cams are molded into one single component, which fastens to the tapped end of the motor output shaft (Figures W9-W10).

a. The inner cam operates the shut-off arm (Figures W10-W11).

c. The center cam operates the holding switch (Figure W10).

d. The front cam operates the water valve (Figure W10).

The positioning of the timing cam is necessary to obtain the correct positioning of the ejector blades at the completion of the harvest cycle.

HOLDING SWITCH (Figure W10)

This is a single pole single throw switch that is operated by a cam and is secured to the motor housing by one screw. This switch assures completion of the harvest cycle after the first few degrees of motor rotation. The switch is operated by a plunger that is depressed most of the time.

CHECKING THE HOLDING SWITCH

a. Remove wires from switch.

b. Remove screw securing switch to housing.

c. Check with a continuity tester.

WATER VALVE SOLENOID SWITCH (Figure W10)

This is also a cam operated switch, single pole, but double throw. This switch also is secured to the motor housing by one screw. Normally closed position will allow the motor to operate through the holding switch. The normally open side energizes the water valve solenoid when the lever arm depresses the switch plunger. A raised spot on the timer cam activates the switch (Figure W10).

SHUT-OFF ARM AND SWITCH

The cam driven shut-off arm operates an electrical switch to control the quantity of the ice crescents produced. During each harvest cycle the arm is raised by a lever that follows the rear portion of the cam. It is then lowered and comes to rest alongside the ice mold. If the bin should be full, the arm remains in the raised position, leaving the shut-off switch open. When enough ice has been removed from the ice bin, the arm will lower, closing the switch, and allowing another harvest cycle to occur.

MOTOR

The motor used is a one direction motor, a stall-type, shaded pole, and it drives a reducing gear and a single output shaft. One end of the output shaft connects to the ejector blades; the other end is coupled to the timing cam (Figure W9, W12).

Unloaded, the shaft makes a revolution in 66 seconds. The motor is replaced as an assembly.

LIMIT SWITCH (Figure W12)

A bi-metallic thermo disk acts as a limit switch. It is in physical contact with the ice mold, and held in place by a spring clip. In the event of extreme high temperatures in the ice mold, the limit switch opens to de-energize all the electric components of the ice maker. The limit switch opens at 135°F. plus or minus 12°, and closes at 95°F. plus or minus 15°, and resets itself automatically.

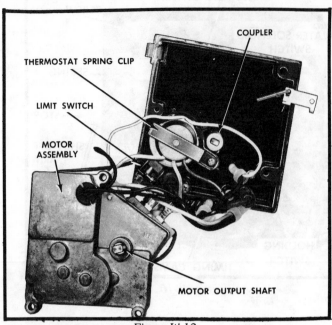

Figure W-12
MOTOR REMOVED FROM SUPPORT

THERMOSTAT AND RESET HEATER CONDUCTOR ASSEMBLY

Bonded to the inside of a die cast heat conductor with a metallic cement, is the ice mold thermostat. This component is a bi-metallic disc type thermal switch, with a sensing surface. Whenever the thermostat or heat conductor is replaced, the parts must be re-bonded with the proper cement (Figure W13).

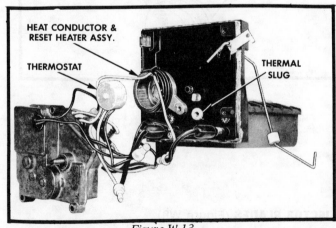

Figure W-13
THERMOSTAT, RESET HEATER AND CONDUCTOR

The heat conductor is in physical contact with the thermal slug which, in turn, makes contact with the front ice piece. In this manner, the temperature of the front piece is transferred to the thermostat. During the time that the water in the ice mold is freezing, the WARM terminal of the thermostat is connected in the ice maker circuit. When the thermostat senses a 16° F. plus or minus 3°, it switches its COLD terminal into the ice maker circuit. This will start the harvest cycle by applying power to the mold heater and motor.

Once the harvest cycle begins, it is necessary to reset to its WARM position to complete the cycle. This occurs when it senses a temperature of 31°F. plus or minus 6°. To accomplish this, a reset heater is used. A ten watt resistance wire is wrapped around the heat conductor (Figure W13). The reset heater and conductor are one assembly which is held by a spring clip (Figure W12).

ELECTRICAL OPERATION

The electrical circuit of the ice maker is not complicated. The automatic ice ejection, water refilling, and controlled storage of the ice are accomplished by the proper functioning and timing of the components previously described. To explain the complete electrical cycle, a series of wiring diagrams are used. Note the position of the ice injector in these diagrams.

FREEZE CYCLE (Figure W14)

In the freezer cycle, the ice mold is filled with water and the shut-off arm has closed the shut-off switch. The mold thermostat is in the warm position and the ice on the ejector blades was produced during the previous harvest cycle. The heater, motor and water solenoid coil are all de-energized.

EJECTION CYCLE

The water in the mold is frozen. When the temperature of the front ice piece is reduced about 16°F., the thermostat switches from warm to cold (Figure W15).

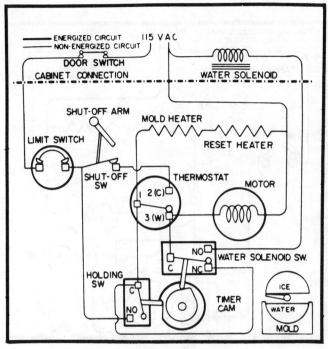

Figure W-14 FREEZE CYCLE

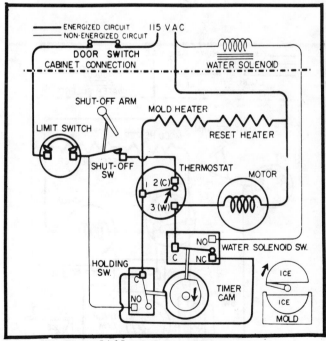

Figure W-15 START OF EJECTION CYCLE

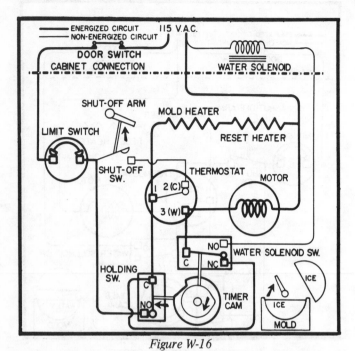

Figure W-16

HOLDING SWITCH CLOSING-SHUT OFF SWITCH OPENING

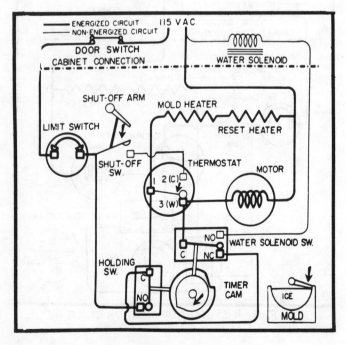

Figure W-17 STALL PERIOD

This energizes the mold and reset heaters. The motor is also energized and its circuit is through the cold terminal of the thermostat to the common terminal of the holding switch, through the normally closed terminal of the water solenoid switch, to the warm terminal of the thermostat and to the motor. The ejector blades and timing cams begin to turn. After a few degrees of the motor shaft rotation, the timing cam closes the holding switch (Figure W16). As the cam continues to rotate, the shut-off arm is raised, opening the shut-off switch. Operation still continues as power is fed to the heaters and the motor through the holding switch (Figure W16). After the holding switch is closed, it remains in this position throughout the ejection cycle to provide power for the heater and motor. As the ejector blades rotate, the ice resting on top of the blades falls into the ice bin (Figure W16).

The ejector blades continue to rotate until they come in contact with the frozen ice crescents. The motor then stalls, as power continues to be applied (Figure W17), but will not damage the motor because of its high impedance. The motor is stalled until the crescents melt loose from the mold, as the heater is still working during this period. From the time the harvest cycle begins and during the time the motor is in a stalled position, the reset heater usually warms the thermostat enough to cause its contacts to switch back to warm position (Figure W17).

When the ice is free from the mold, it is swept out of the mold by the ejector blades (Figure W18). During this time, the rotating cam allows the shut-off arm to lower, closing the shut-off switch, unless the bin is already full. The cam and ejector blades continue to rotate and at the proper time, the timing cam disconnects the normally closed terminal of the water solenoid switch and connects the normally open terminal to the warm terminal of the thermostat. The solenoid coil is energized through the holding switch and thermostat (Figure W18). This opens the water valve, permitting the water to refill the mold. This takes about 7 or 8 seconds, at which the cam advances, enabling the water solenoid switch to return to its normally closed position. The solenoid coil is de-energized at this time and closes the water valve. The heaters are still in the circuit and the motor is stalled, but pressing through the holding switch and thermostat.

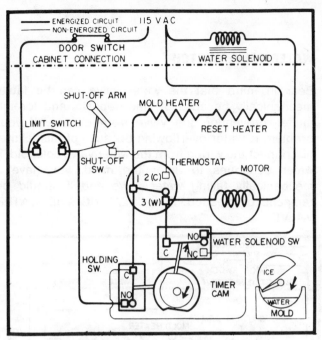

Figure W-18 WATER SOLENOID ENERGIZED

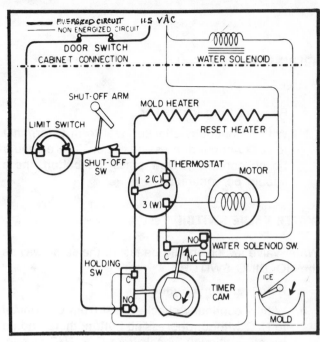

Figure W-19 THERMOSTAT DID NOT RESET

If the thermostat did not reset from its cold to warm position, the motor would continue to operate until the water solenoid switch changes from normally closed to its normally open terminal. At the same time the motor will be de-energized and power will not be applied to the water solenoid coil (Figure W16). The heater stays on. When the thermostat resets, operation continues.

The motor runs as the ejector blades and the timing cam rotate until the holding switch is open (Figure W17). This is the end of the harvest cycle and the start of another freezing period. The mold refills with water, the ejected ice crescents are atop the ejector blades and power to the heaters and the motor is halted.

The harvest will continue until the bin is full and prevents the shut-off arm from returning the switch to the closed position (Figure W18). The harvest cycle will be completed, however, the open shut-off switch will prevent a new cycle, even when the thermostat switches over to the cold position. Enough ice must be removed to start the operation of the ice maker once more.

TESTING AND ADJUSTING THE ICE MAKER

NOTE: Do not remove ice maker from cabinet.

a. Turn off water supply. Remove any water that has not been frozen.

b. Depress freezer door switch.

c. Slowly turn the ejector blades manually, clockwise, about 30 degrees, so the holding switch is closed (Figure W19). Allow two cycles. These two cycles will allow the heater to warm the cold metal components and prevent condensation on the electric components.

CAUTION: Be careful that fingers do not get caught in ejection blades after rotation starts.

d. Remove the front cover and again repeat the ejection cycle as in line (c); check the operation of the electric components. Once the motor starts running, all components can be visually inspected for operation, except the thermostats. Repeat this operation if necessary.

e. If there is obviously no mechanical failure, the switches and the heater and thermostats will have to be checked individually.

f. If an ohmmeter is not available, a test cord can be made, (Figure W20).

HOLDING SWITCH, Test

   a. Remove from ice maker.

   b. With a continuity checker, test from C terminal to N/O without depressing button. There should be no continuity. Now press the button, there should be continuity.

WATER VALVE SWITCH, Test

Water valve switch should be tested the same way as the HOLDING SWITCH.

   a. Remove from ice maker.

   b. With a continuity checker, test from C terminal to N/O without depressing button, there should be no continuity, check from C to N/C closed, still without depressing the button, it should show continuity.

   c. Now depress the button, from C to N/O, there should be continuity, from C to N/C, there should be no continuity.

   d. If either of these tests fail, replace the switch.

ADJUSTING THE HOLDING SWITCH

   a. Position the cam notch, as shown in Figure W21, by turning the ejector blades. This should position the holding switch plunger just below the cam notch.

   b. Loosen the switch mounting screw and move the switch so that the tip of the plunger just reaches the flat of the center cam by the cam notch and is not depressed.

   c. Tighten the mounting screws and test cycle the ice maker.

POSITIONING THE EJECTOR BLADE

Completing the cycle, the ejector blade should come to rest when it is even with, or ¼" above, the left side of the mold (Figure W22).

   a. Manually position the ejector blade, as shown in Figure W22.

   b. Loosen the cam mounting screw and adjust the position of the cam so that the holding switch opens (plunger released). Tighten the cam mounting screw. Test cycle the ice maker, consider the cycle completed when the holding switch opens.

WATER VALVE SWITCH

Bear in mind that the water valve can be faulty, mechanically, because of lime deposits and foreign matter preventing the valve from closing. If your problem is water overflowing because of this failure, disconnect the wiring from the water valve solenoid. If water continues to enter the mold, you have a mechanically failing water valve and it should be replaced or cleaned. See CLEANING THE WATER VALVE.

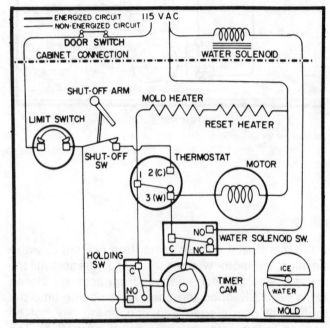

Figure W-20 END OF EJECTION CYCLE

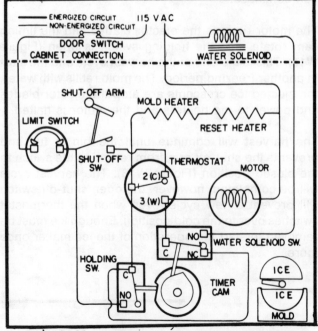

Figure W-21 ICE BIN FULL—SHUT OFF SWITCH OPEN

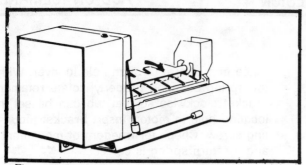

Figure W-22 MANUALLY STARTING EJECTION CYCLE

ADJUSTING THE WATER VALVE SWITCH

a. Test cycle the ice maker, measure the volume of water. If volume is too much or not enough, proceed with the following:

b. Loosen the switch mounting screw. Raise the switch to decrease or lower the switch to increase the volume.

c. Tighten the switch mounting screw, test cycle the ice maker.

d. It may be necessary to repeat this operation, until the water volume is correct.

SHUT-OFF ARM SWITCH

The proper bin capacity will be maintained when the switch closes at the time the bottom of the shut-off arm is ¾" from the mold support (Figure W21).

a. Loosen the switch mounting screw.

b. Move the switch to the left to increase the spacing, or to the right to decrease the spacing.

c. Recycle and check spacing.

PARTS REPLACEMENT PROCEDURES (Figure W23)

Disconnect all power to the ice maker. Allow ice maker to warm to room temperature to prevent condensation of electrical items.

FILL TROUGH AND BEARING

a. Pull back tab at side of trough, twist in clockwise direction.

b. Pull from back to detach from the ejector blade shaft.

c. Replace new parts in reverse order.

EJECTOR BLADES

a. Place blades in the twelve o'clock position.

b. Remove the fill trough and bearing.

c. Move blades back and upward to disengage from the motor coupling.

d. Place a small amount of silicone grease on bearing ends of replacement ejector.

e. Replace in reverse order, being sure that blades fit in motor couple at the twelve o'clock position.

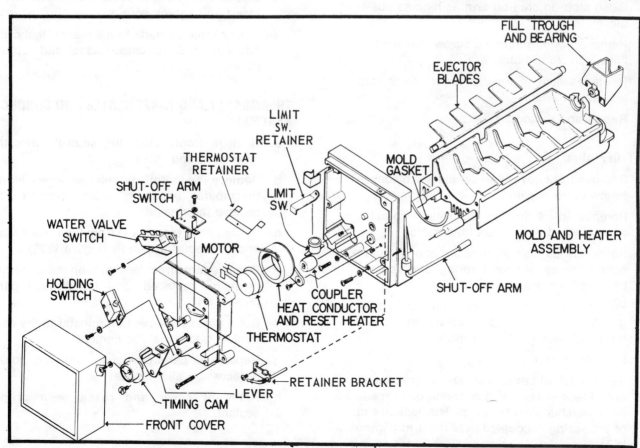

Figure W-23 PARTS BREAKDOWN

FRONT COVER

a. Remove mounting screws from top and bottom of cover, pull straight away from ice mold.

b. Clean sealing compound from inside of cover and dry mechanism area completely.

c. Place a sealer around mold support to seal cover and wiring, mounting clip and heater. *NOTE: Always use a sealer when cover is to be removed and replaced.*

d. Replace cover, remove excess sealer, wipe surface clean, wipe cover and mold support.

WATER VALVE SWITCH AND HOLDING SWITCH

a. Remove front cover, disconnect wires from switch terminals.

b. Remove switch mounting screws, pull switch from mounting peg on motor.

c. Replace switch in reverse order, recycle and test for proper adjustment.

d. Reseal and mount front cover as previous.

SHUT-OFF ARM SWITCH

a. Remove cover, disconnect the wires.

b. Raise stop on shut-off arm as high as possible.

c. Remove switch mounting screw and washer, lift switch from motor.

d. Replace switch in reverse order, lower stop and adjust switch.

e. Reseal and remount cover.

SHUT-OFF ARM

a. Remove front cover and set ejector blades in freeze cycle position.

b. Remove screw holding arm retainer bracket to motor and lift bracket from motor.

c. Raise arm as high as possible and lift out of hole in lever. Remove spring and retainer bracket from arm.

d. Pull arm back through support as far as it will go. Carefully bend arm down and slip it through and out hole in support.

e. Replace in reverse manner.

f. Insert notched tab side of retainer bracket on arm. Place straight end of spring on arm and set in notched tab of retainer. The hooked end of the spring is coupled over the arm (Figure W24).

g. Place end of arm through hole in lever. With the spring engaged properly, rotate retainer bracket clockwise so that tab can be set in locating hole in motor. Insert bracket mounting screw. Check for freedom of movement and if return spring is correct.

h. Reseal and replace cover.

MOTOR

a. Remove the cover of the ice maker. Remove the holding switch, the water solenoid switch and the shut-off arm switch.

b. Remove the timing cam, shut-off arm and lever arm.

c. Remove the three motor screws that hold the motor to the support, and remove the motor shaft from the ejector blade coupler.

d. Disconnect motor wires from terminals, and reconnect new motor.

e. Align motor with mounting holes and with coupler on ejector blade shaft. Rotate ejector blades until coupler locks into motor shaft. Insert and tighten screws.

f. Reassemble all parts to ice maker, test cycle and adjust all switches. Reseal and replace cover.

THERMOSTAT AND RESET HEATER AND CONDUCTOR ASSEMBLY

a. Remove front cover, the shut-off arm, the lever arm, and motor.

b. Remove the thermostat retainer screw, lift off thermostat assembly from support and remove the retainer.

d. Separate the thermostat from the reset heater and conductor assembly (Figure W23).

e. Remove wires from terminals on the parts that must be replaced, and install new parts (Figure C16).

f. Bond the replacement thermostat to the conductor with a metallic cement.

g. Replace all parts previously removed, adjust if necessary, all switches.

h. Remount cover and reseal with proper sealant.

MOLD AND HEATER ASSEMBLY

a. Remove the front cover as previous.

b. Remove the following parts: the fill trough, the ejector blades, the shut-off arm, the motor, the thermostat and the heater assembly and limit switch.

c. Remove inoperative mold heater wiring leads from the connectors (Figures W25-W26).

d. Remove four screws that secure the mold and heater assembly to the mold support, one of the screws holds the limit switch retainer

e. Remove the mold by pulling the heater leads through the mold support.

f. Replace the mold support gaskets, packaged with mold and heater assembly.

g. Replace mold and heater assembly. Replace all parts and adjust if necessary.

h. Seal mold heater at mold support with per-magum.

i. Cover limit switch terminal near support with insulated sleeve.

j. Be sure timing cam and ejector blades are both at start position.

k. Reseal and remount front cover.

LIMIT SWITCH

a. Remove front cover, the shut-off arm, the lever arm, the motor, the thermostat and reset heater assembly.

b. Loosen and remove the screw from the limit switch retainer, remove switch.

c. Disconnect wires from inoperative switch, reconnect new switch, and mount.

d. Replace all parts, adjust if necessary, reseal and replace cover.

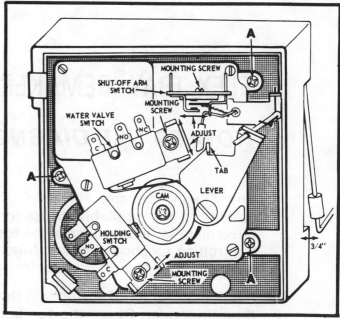

Figure W-24 SWITCH ADJUSTMENTS

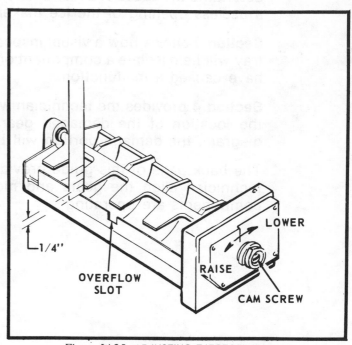

Figure W-25 ADJUSTING EJECTOR BLADES

WHIRLPOOL DESIGN

FLEX TRAY ICEMAKER

HOW TO USE DIAGNOSTIC & REPAIR GUIDE

The Diagnostic and Repair Guide is designed to help the technician determine the probable cause or causes of a malfunction BEFORE the coverplate is removed from the ice maker. Since the flex tray is one of the controlling functions of the refrigerator, the technician must determine if the problem is in the refrigerator or in the ice maker.

Section 1 of the guide will help the technician determine if the problem is in the refrigerator. Check the components and/or systems before checking the ice maker.

Section 2 shows how a visual inspection of the ice in the bucket can help determine the cause of a malfunction. This vital visual inspection may prevent a useless opening of the ice maker and possibly a second service call.

Section 3 shows how a visual inspection of the condition and position of the tray will help isolate a component or condition within the ice maker that may have caused a malfunction.

Section 4 provides the technician with possible causes of stripped gears by the location of the damaged gear teeth. Simply place the gear over the diagram, the damage portion will be in area 1, 2, or 3.

The back page of this guide provides diagnostic and repair tips to help the technician align the gears and motor pinion and quickly check if some components are operating.

TROUBLE DIAGNOSIS AND CHECKING PROCEDURE

★ To align motor pinion, use Quick Check or T.V. cheater cord.

★ Plug cheater cord into the center pins (pink & red). See Figure 1. Operate motor until a pinion gear tooth is aligned with the pointer in the Housing. See Figure 2.

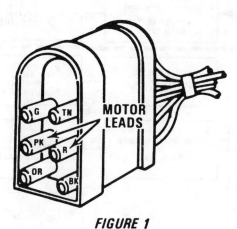

FIGURE 1

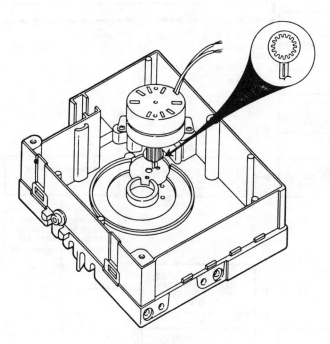

FIGURE 2

★ For more positive gear alignment, use the small end of the drive pin as an alignment tool. See Figure 3.

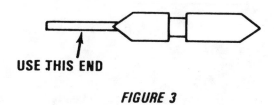

USE THIS END

FIGURE 3

★ For positive alignment of the drive gear use drive pin and center portion of a cam Cut cam as shown in Figure 4.

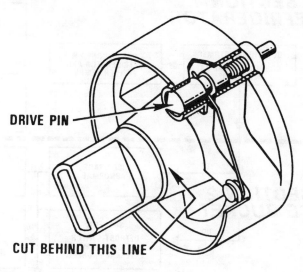

DRIVE PIN

CUT BEHIND THIS LINE

FIGURE 4

★ Quick check. Turn Refrigerator Thermostat to coldest position. Using screwdriver as a sound transmiter, listen for 60-cycle hum. This proves the motor is energized and the defrost bimetal is closed. If the fans are running, it also proves:

The Defrost switch is in the freeze position.
The thermostat is closed.

★ If the 60-cycle hum is present and the fans are not running, the switch is in the defrost position.

★ If nothing operates (Fans or Compressor) and the 60-cycle hum is not present, check the defrost indicator. If the marks are aligned as in Figure 5, check for defrost switch deadbreak. Deadbreak occurs when the defrost switch is not closed in defrost or refrigeration. See Figure 6.

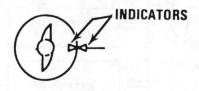

INDICATORS

FIGURE 5

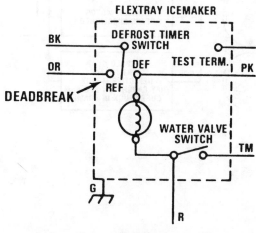

FLEXTRAY ICEMAKER

FIGURE 6

SECTION 1
REFRIGERATOR

SECTION 2
ICE BUCKET

SECTION 3
ICE TRAY

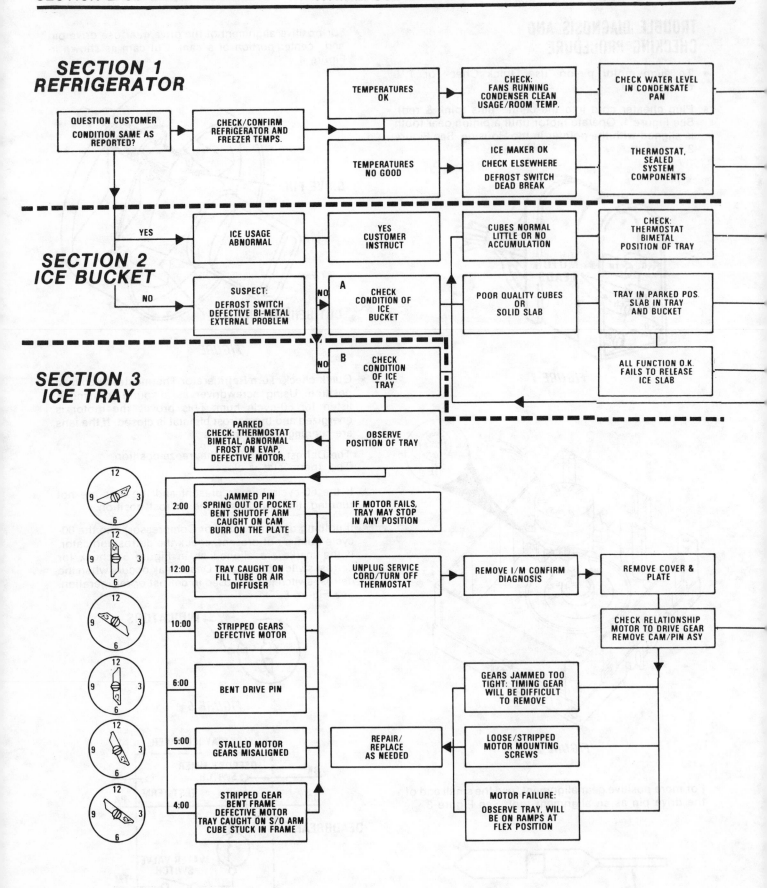

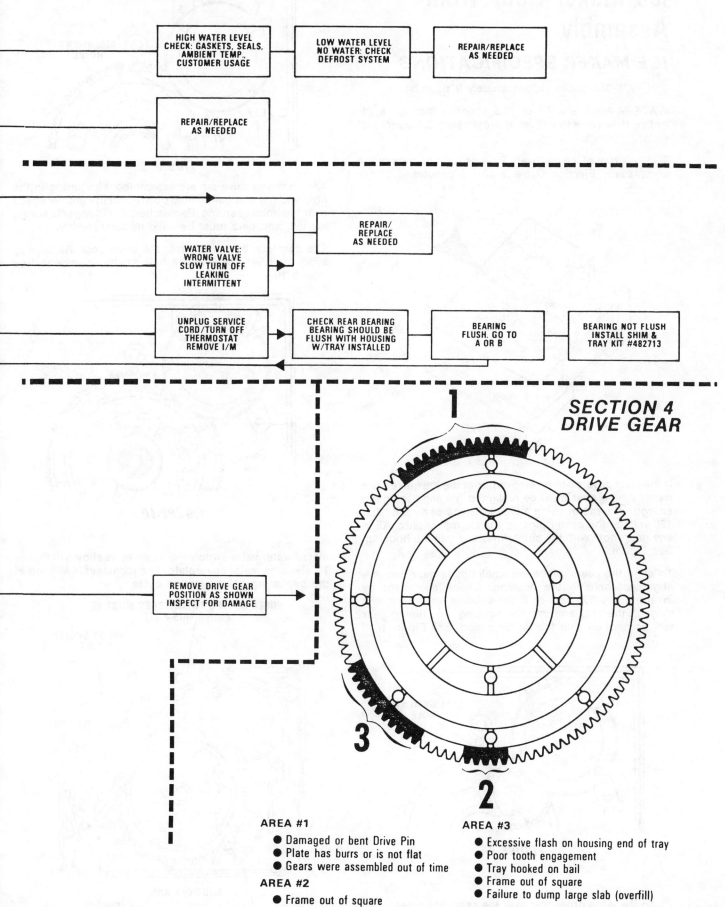

HIGH WATER LEVEL
CHECK: GASKETS, SEALS,
AMBIENT TEMP.,
CUSTOMER USAGE

LOW WATER LEVEL
NO WATER: CHECK
DEFROST SYSTEM

REPAIR/REPLACE
AS NEEDED

REPAIR/REPLACE
AS NEEDED

WATER VALVE:
WRONG VALVE
SLOW TURN OFF
LEAKING
INTERMITTENT

REPAIR/
REPLACE
AS NEEDED

UNPLUG SERVICE
CORD/TURN OFF
THERMOSTAT
REMOVE I/M

CHECK REAR BEARING
BEARING SHOULD BE
FLUSH WITH HOUSING
W/TRAY INSTALLED

BEARING
FLUSH. GO TO
A OR B

BEARING NOT FLUSH
INSTALL SHIM &
TRAY KIT #482713

SECTION 4
DRIVE GEAR

1

3

2

REMOVE DRIVE GEAR
POSITION AS SHOWN
INSPECT FOR DAMAGE

AREA #1
- Damaged or bent Drive Pin
- Plate has burrs or is not flat
- Gears were assembled out of time

AREA #2
- Frame out of square

AREA #3
- Excessive flash on housing end of tray
- Poor tooth engagement
- Tray hooked on bail
- Frame out of square
- Failure to dump large slab (overfill)

Ice Maker Gear Train Assembly

ICE MAKER SPECIFICATIONS

BIN—275-325 cubes (approximately 9 pounds).

WATER—Automatic fill of 12.5 seconds through a tri-orifice flow valve to deliver approximately 6.5 ounces of water.

TIME—1 tray of ice per every 2 hours running time of the compressor. Ejection Cycle is 13-1/3 minutes.

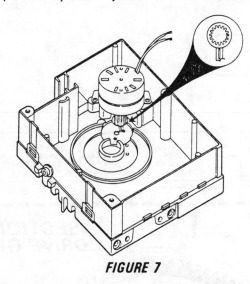

FIGURE 7

If the steel coverplate is removed from the ice maker, the motor pinion gear must be realigned. To align the gear, energize the motor using a Quick-Check or by applying 120 volts to the center pins on the ice maker plug. Align any gear tooth with the pointer molded into the housing. See Figure 7.

To align the gears, place the small timing gear over the alignment hole in the housing. Check alignment by inserting a 3/32" rod or other suitable alignment tool through the gear and into the housing. The timing gear will be meshed with the motor pinion. See Figure 8.

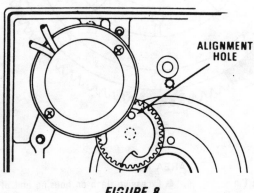

FIGURE 8

28 Place the large drive gear over the same alignment hole and reinsert the alignment rod. The larger center hole MUST be centered over the hole in housing. See Figure 9.

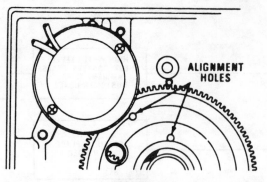

FIGURE 9

The cam and drive pin are assembled with spring in the hole and the plunger pushed down until the groove snaps over the holding spring. Remember, this plunger is spring loaded, and care must be used in disassembly.

The cam can be placed over the drive gear. Be sure to align the drive pin with the hole in the drive gear. See Figure 10.

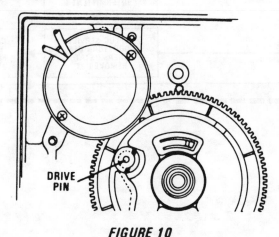

FIGURE 10

Install water valve switch. Route wires as shown in Figure 11. Replace metal coverplate. The icemaker is aligned at the beginning of an ejection cycle.

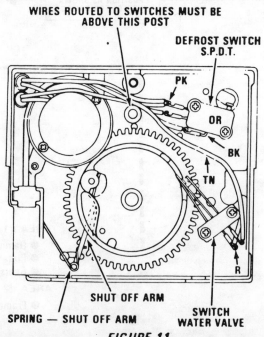

FIGURE 11

DEFROST GEAR TRAIN ASSEMBLY

The defrost gears are on the tray side of the control housing, and the cover is secured by three screws. As with the other gears, once the cover is removed, the entire gear train must be reassembled.

The first step in assembling the defrost gear train is very important. The gear must be placed on the control housing so that the bosses are toward housing, and one of the cam openings is over the pilot hole. Insert a 3/32" rod or other alignment tool. See Figure 12.

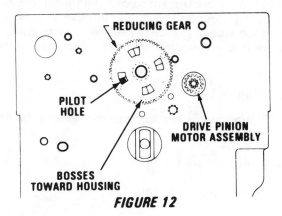

FIGURE 12

With the alignment tool in the pilot hole, position the cam and pinion as shown in Figure 13.

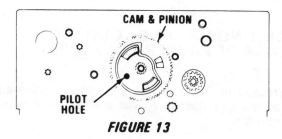

FIGURE 13

Install the cam follower lever as shown in Figure 14.

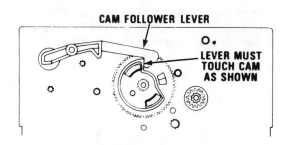

FIGURE 14

The gear and pinion now must be assembled with the pinion engaging the cam gear and the timer motor gear. Pinion is down or toward the housing. See Figure 15. Alignment of this gear is not necessary.

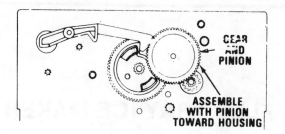

FIGURE 15

Next, install the large cam and gear over the alignment rod. The cam must engage the cam lever as shown in Figure 16.

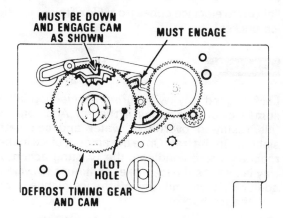

FIGURE 16

With the alignment tool in place, install the cover. The cover is also the bearings for the gears. Install the three screws and remove the alignment tool.

The defrost gear train is now assembled and it can be advanced by inserting a coin in the "S" slot on the large cam and gear, and rotating clockwise. See Figure 17. **Do not install icemaker in defrost position.**

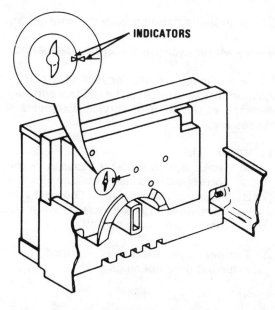

FIGURE 17

FLEX TRAY ICE MAKER

ICE SLABBING INTO BUCKET

PROBLEM:

Failure to eject ice cubes from Flex Tray — ice slab — ice bucket.

CAUSE:

Lack of good tray twist during ice ejection, allowing some ice cubes to remain in the tray. Additional water fills result in overfill of water and complete ice slabbing in the tray. Once the amount of ice in the tray reaches a slab condition, the twisting action of the tray is lost. As the ice maker continues to cycle, each additional water fill runs off the ice-slabbed tray and into the ice bucket.

SERVICE TECHNICIANS OBSERVATIONS:

1. Ice Maker tray is in the parked position and completely full of ice.

2. Ice Bucket will have a reasonably level slab of ice in the bottom of the bucket. (If no ice is present, check with customer. May have emptied slab prior to service call).

3. Ice droplets hanging from the shut-off arm.

4. Ice Maker cycles and controls defrost normally.

There are many causes for lack of a complete ice ejection. These causes can be random and may not be detected at the time of service. Examples of such causes are:

1. Water Valve:
   A. Leaks due to mineral deposits — may self-correct with additional water fills.
   B. Incorrect water valve — Compact valve used by mistake. This will increase water flow rate.
   C. Slow water valve turnoff.

2. Temperature increases caused by usage, extended door openings.

DIAGNOSIS AND CORRECTION — MECHANICAL

Mechanical malfunctions can cause a failure to eject ice from the tray. These malfunctions usually result in stalled trays and customer complaints of little or no ice in the ice bucket. These complaints can easily be separated from random causes by observing the **lack of an ice slab** in the ice bucket. When a mechanical failure is diagnosed, the cause is most likely related to:

1. Stripped gears

2. Slotted motor too loose or too tight to the drive gears

3. Out-of-square frame

4. Misalignment of gears

5. Damaged drive pin

6. Motor failure

Repair defect.

DIAGNOSIS AND CORRECTION — LACK OF PROPER TRAY TWIST

Proceed to diagnosis and perform corrective action for lack of proper tray twist and failure to eject ice from tray:

1. Unplug refrigerator and remove the ice maker.

2. Observe the tray-to-rear-bearing-assembly area located at the rear of the frame, *Figure 18*.

3. With ice tray in a parked position, the tray should hold/push the spring-loaded bearing flush with the bearing assembly, *Figure 18*.

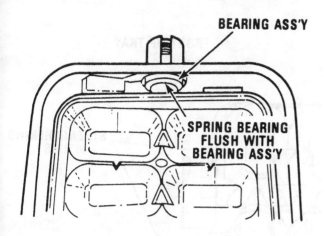

FIGURE 18

**NORMAL ENGAGEMENT — SPRING BEARING
FLUSH WITH BEARING ASS'Y
WITH TRAY IN "PARK" POSITION**

4. If the spring-loaded bearing protrudes out from
the bearing assembly, use the shims supplied in
Kit #482713 to space the rear bearing assembly
out from the metal frame. (By using the shim
thickness as a gauge, the amount of shimming
required can be judged by placing the shim at the
point where the tray and spring-loaded bearing
meet). This will give a good indication of how
many shims are required. **No more than two
shims should be used per ice maker**, Figure 19
and Figure 20.

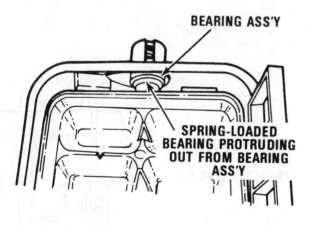

FIGURE 19

**POOR ENGAGEMENT — SPRING BEARING
PROTRUDES OUT FROM BEARING ASS'Y
WITH TRAY IN "PARK" POSITION**

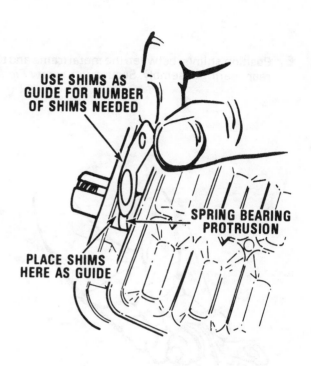

FIGURE 20

5. Remove the tray and disassemble the rear
bearing assembly from the metal frame, Figure
21.

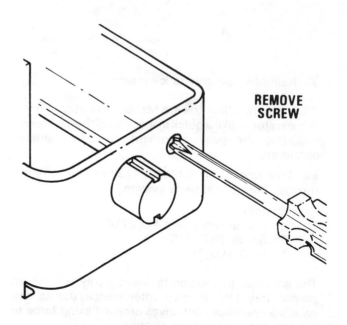

FIGURE 21

FRONT OF TRAY

6. Position shim(s) between the metal frame and the rear bearing assembly. Secure with screw *Figure 22* .

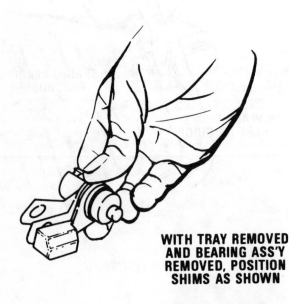

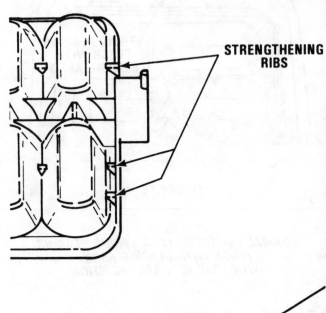

STRENGTHENING RIBS

WITH TRAY REMOVED
AND BEARING ASS'Y
REMOVED, POSITION
SHIMS AS SHOWN

ASSEMBLE SHIMS
AND BEARING ASS'Y
TO FRAME AND
REINSTALL TRAY

FIGURE 22

NEW TRAY IMPROVEMENTS

REAR OF TRAY

7. Reinstall new tray to ice maker.

Prior to reinstalling the Ice Maker Assembly into the refrigerator, make a quick inspection of the Ice Maker, including the gear alignment, to insure proper operation.

Ice Tray and Shim Kit #482713 may be ordered through the normal parts system.

Kit contains:
   1   Instruction Sheet #482714
   2   Shims #482715
   1   Ice Maker Tray

The shims(s), in addition to the new tray, will insure proper tray tab, to stop interference during the twisting operation, and gives proper flexing force to the tray for complete ice ejection.

See tray changes in *Figure 23* .

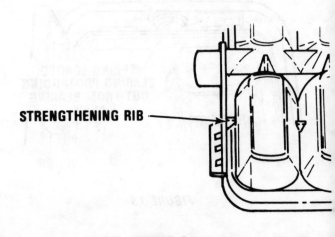

STRENGTHENING RIB

FIGURE 23

WHIRLPOOL DESIGN

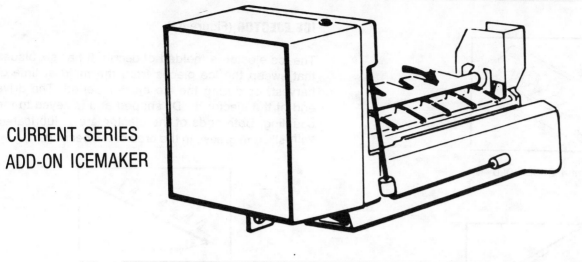

CURRENT SERIES ADD-ON ICEMAKER

NORMAL SOUND CHARACTERISTICS

If you have never had an automatic ice maker in your refrigerator before, you may be alarmed at the various noises and sounds that are produced by the ice maker mechanism.

1. The motor will hum when it is in the stalled position during the ice harvest.
2. The breaking away of the ice from the molds.
3. The ice falling into the storage bin below.
4. The running of water and the sudden snap of the water valve as it closes.
5. The sound of dripping water as the mold heater defrosts the cubes prior to the harvest.
6. The ejector blades not completing its cycle due to ice sticking in the ice mold, this will be an intermittent sound prior to the harvest.
7. The ice maker shutting off upon opening the freezer door.

All of these sounds may be slightly annoying at first but they are built in to the ice maker and should not cause any worry or undue anxiety, Figure A1.

LENGTH OF CYCLE

The harvest of ice will vary from 45 minutes to 2 hours. The time is effected by numerous things.

   a. How many times you may open the door.

   b. The temperature of the ambient air in the room.

   c. The water temperature entering the ice mold.

   d. The load condition of the refrigerator.

Because of this, it is hard to make a comparison even with the identical model and make of refrigerator. If there is a specific complaint of no ice, or not enough ice, then there will be a need to refer to the diagnosis charts of this Repair-Master.

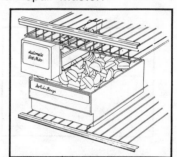

Figure A-1 AUTOMATIC ICE MAKER

COMPONENT DESCRIPTION AND CHARACTERISTICS
ICE MOLD

The ice mold is a die cast aluminum unit that will mold six ice crescents, later models of this same unit will mold 8 or 12 crescents. The front surface contains a thermostat that is bonded to this surface. A film of silicone grease is placed on the top edge of the mold to prevent siphoning of the water through a capillary action, Figure A2. The mold heater is rated at 165 watts; it is used to thaw the ice from the mold. The heater is in series with the thermostat that is on the front surface of the mold. This thermostat closes with drop of temperature and is normally open. The original heater is staked in place to the bottom of the mold. With the use of four flat head screws a replacement heater can be installed upon failure of the original heater (Figure A3).

ICE STRIPPER

The ice stripper is made of plastic and attached to the dumping side of the mold. It is used as a decorative piece and the protruding fingers prevent the ice pieces from falling back into the mold, Figure A4.

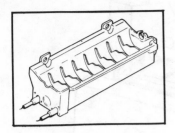

Figure A-2 ICE MOLD

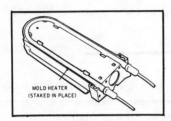

Figure A-3 MOLD HEATER (STAKED IN PLACE)

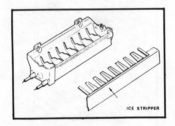

Figure A-4 ICE STRIPPER

THERMOSTAT

The thermostat is a single-pole, single-throw, bi-metal switch. It closes upon drop of temperature and is normally open. The thermostat closes at 12° plus or minus 3°F., and will reset itself at 48° plus or minus 6°F. It is wired into the circuit in series with the heater and will act as a safety device against overheating in the case of mechanical failure of the ice maker (Figure A5).

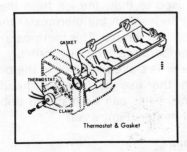

Figure A-5 THERMOSTAT AND GASKET

ICE EJECTOR, (Figure A6).

The ice ejector is molded of decrin. It has six blades that sweep the ice pieces from the mold at time of harvest, or during the ice ejector period. The drive end of the ejector is "D" shaped and is keyed to the coupling. Both ends of the ejector are lubricated with silicone grease in the area of the bearings

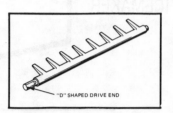

Figure A-6 ICE EJECTOR

SHUT OFF ARM AND LINKAGE

The shut-off arm is driven by a cam and operates a switch to control the quantity of the ice produced. During each ejection cycle the shut-off arm is raised and lowered during each of the two revolutions of the timing cam. If the shut-off arm comes to a stop on top of the ice cubes in the storage bin during either revolution, it will automatically stop the ice maker. The shut-off arm can also be controlled manually by raising it to the extreme top, or lowering it will turn the ice maker off or on. (Door must be closed for ice maker to operate.)

TIMING SWITCHES

There are three switches used on this ice maker. They are all of the single pole double-throw type. Although the switches are identical and can be interchanged for service, they function differently.

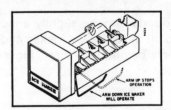

Figure A-7. SHUT-OFF ARM

1. HOLDING SWITCH, this switch will assure a complete revolution once the ice maker has started.
2. WATER VALVE SWITCH, controls the flow of water during the fill period. This switch can be adjusted.
3. SHUT-OFF SWITCH, the shut-off switch is activated by the shut-off arm and linkage. Its purpose is to stop the operation of the ice-maker (Figure A8).

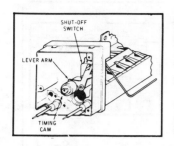

Figure A-8 -SHUT-SWITCH

TIMING CAM AND COUPLER

There are three cams combined into the molded delrin component. One end is attached to a large timing gear, the other end is coupled to the ejector. Each cam controls certain switches.

a. The inside cam controls the shut-off switch, the arm and linkage.

b. The middle cam operates the holding switch.

c. The outside cam controls the water valve switch (Figure A9).

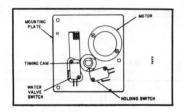

Figure A-9 HOLDING AND WATER VALVE SWITCH

TIMING GEAR

The timing gear is driven by the pinion gear on the motor and rotates the cam and ejector. The cam is keyed with a "D" shaped protrusion to a mating "D" shaped hole in the timing gear. The gear must be properly spaced to prevent binding against the mounting plate (Figure A10).

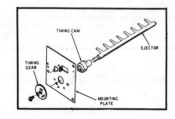

Figure A-10 - TIMING GEAR AND CAM

MOTOR (Figure A11)

The motor used is a low wattage, stall type motor. It has a single output shaft. This shaft mounts a pinion gear which drives the timing gear. The timing cam and ejector rotates at about one revolution every three minutes.

Copy -1

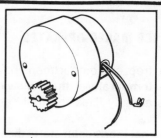

Figure A-11 MOTOR

WATER VALVE

A conventional water valve is used, and a flow washer controls the water pressure. There is a strainer in the water valve that should be cleaned occasionally. The solenoid coil of the water valve is connected in series with mold heater.

Figure A-12 -WATER VALVE

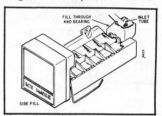

Figure A-13   FILL TROUGH AND BEARING

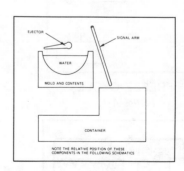

Figure A-14

FILL TROUGH AND BEARING

This part is molded nylon. It is used to support the inlet tube and directs the water filling the mold. It also acts as a bearing to support the far end of the ejector blade (Figure G13).

WIRING

HUNTINGTON CITY-TOWNSHIP
PUBLIC LIBRARY
200 West Market Street
Huntington, IN 46750

When working on the ice maker it is important to place all wiring back in the place where it is removed from to prevent tangling with the mechanism. A wiring diagram for the ice maker is inside the front cover.

HOW THE ICE MAKER OPERATES

When the freezer compartment reaches its normal operating temperature the following takes place. Water has already filled the mold and is frozen.

1. The internal thermostat starts the ice maker motor.

2. The heater and the ejector blades are energized.

3. The shut-off arm is raised.

4. The ejector blades rotate until they are stopped by the ice in the mold. The motor remains in this stalled position, still energized, until the ice breaks loose from the mold.

5. The heater, meanwhile, is on and heats the mold to allow the ice to break free. The ejector will make a second revolution after the ice is ejected, and the mold heater remains energized.

6. The stop arm raises while the ice pieces are being dropped, then lowers to the level of the ice in the storage bin. When the storage bin gets full, the stop arm will cut off operation of the ice maker.

7. As the ejector blades return to their original position, the thermostat will reset, and the mold will once again fill with water. Follow the Figures A14 thru A24 for cycle of operation.

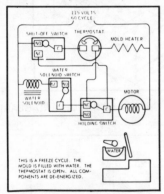

FREEZE CYCLE
Figure A-15

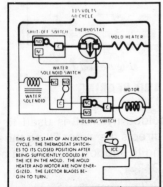

START OF EJECTION CYCLE
Figure A-16

ADJUSTING THE WATER VALVE (Figure G27)

The quantity of water that enters the ice mold can be adjusted. Because there are many different manufactured valves that are used, there will be slight

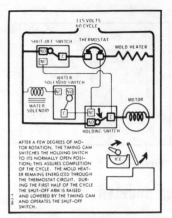

HOLDING SWITCH CLOSING-SHUT OFF SWITCH OPENING   *Figure A-17*

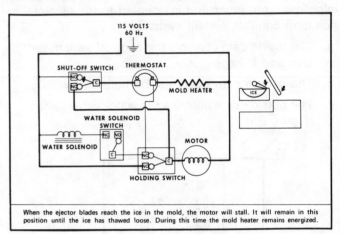

When the ejector blades reach the ice in the mold, the motor will stall. It will remain in this position until the ice has thawed loose. During this time the mold heater remains energized.

STALL PERIOD *Figure A-18*

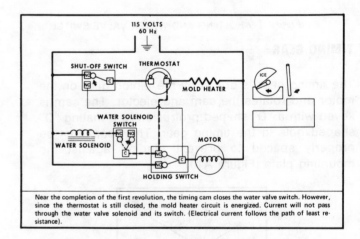

Near the completion of the first revolution, the timing cam closes the water valve switch. However, since the thermostat is still closed, the mold heater circuit is energized. Current will not pass through the water valve solenoid and its switch. (Electrical current follows the path of least resistance).

END OF EJECTION CYCLE *Figure A-19*

variations in the amount of water entering the mold over a predetermined time. Each valve and ice maker is matched and tested before it leaves the factory, however, should it be necessary to adjust the valve, or if a new valve is installed, this valve must be adjusted. To check the water fill proceed as follows:

1. Place a cup under the water fill tube.

2. Remove the plastic cover of the ice maker.

3. Manually cycle the ice maker and collect the water in the cup or any other container.

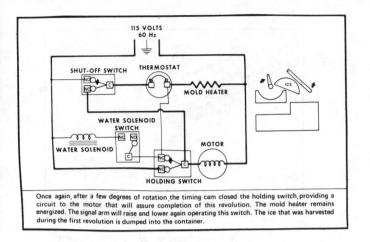

Once again, after a few degrees of rotation, the timing cam closed the holding switch, providing a circuit to the motor that will assure completion of this revolution. The mold heater remains energized. The signal arm will raise and lower again operating this switch. The ice that was harvested during the first revolution is dumped into the container.

Figure A-20

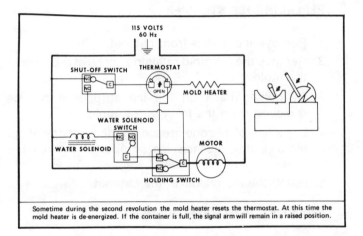

Sometime during the second revolution the mold heater resets the thermostat. At this time the mold heater is de-energized. If the container is full, the signal arm will remain in a raised position.

Figure A-21

4. Measure the water collected in a graded container, such as a baby bottle or any vessel graduated in cc's.
5. The correct amount of water will be 145 cc's.
6. Determine if you are short water, or too much water.
7. Turning the adjusting screw clockwise will decrease the water fill. Counter clockwise will increase the water fill.
8. One complete revolution of the adjustment screw will vary the fill approximately 18 cc's (Figure A27).
9. Repeat cycling and adjusting until the water is right.

REMOVING THE ICE MAKER

1. Remove the three mounting screws from the side of the freezer liner.
2. Disconnect the wiring at the connector plug.

PARTS REPLACEMENT

Figure A28 illustrates the relative positions of the parts and components. In addition to the regular tools, the service man should have the following:

a. Wire nuts for electrical splicing.
b. Alumilastic for thermostat bonding and heater replacement.
c. Silicone grease for use where specified.

FILL TROUGH REPLACEMENT

1. Remove the ice maker from the cabinet.
2. The retaining tab should be pushed back away from the mold, and rotate the fill trough counterclockwise. Pull the trough straight out the back.
3. Installation can be made in the reverse order. Reinstall and connect ice maker, check water level.

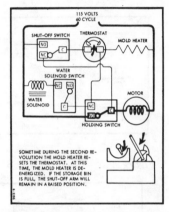

EJECTING ICE INTO BIN

Figure A-22

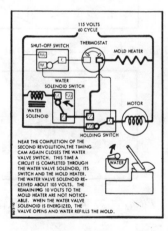

WATER SOLENOID ENERGIZED

Figure A-23

CHECKING THE ICE MAKER

To test cycle an icemaker and check its operation, you will have to make up a test cord, Figure A25.

With the test cord the ice maker can be checked, either on a work bench, or while still in the cabinet.

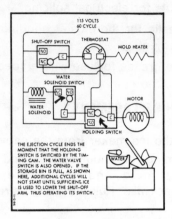

ICE BIN FULL
Figure A-24

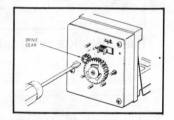

Figure A-25 ICE MAKER TEST CORD

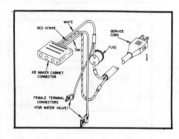

Figure A-26 TEST—CYCLING THE ICE MAKER

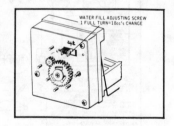

Figure A-27 ADJUSTING WATER FILL

EJECTOR BLADE REPLACEMENT

1. Remove the ice maker from the cabinet.
2. Remove the fill trough and bearing by pushing the retainer tab back away from the ice mold. Rotate the trough counterclockwise and pull straight out the back.
3. Lift the ejector blade up, and slide back to detach it from the front bearing.
4. Use silicone grease on the two bearing ends of the replacement ejector blade, install in the reverse order.
5. Reinstall ice maker, check water level.

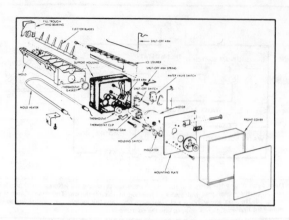

Figure A-28 EXPLODED VIEW OF ICE MAKER

REPLACING ICE STRIPPER

1. Remove ice maker from cabinet.
2. Remove the ice mold retaining screw at the back of the mold.
3. By pulling back on the ice stripper it can be removed from the front of the mold.
4. Lay a film of silicone grease on the top edge of the mold, install the new ice stripper in the reverse order.
5. Install the ice maker in the cabinet.

MOUNTING PLATE REPLACEMENT

1. Remove the ice maker from the cabinet.
2. The front plastic cover will pry loose, start at the bottom of the support.
3. Remove the three screws that secure the mounting plate.
4. Using caution, remove the mounting plate from the shut-off arm and take note as to the relative position, as shut-off arm spring, it MUST be reassembled in the same position.
5. Secure all wiring, make sure it is away from the mechanism.
6. Reinstall in reverse order.

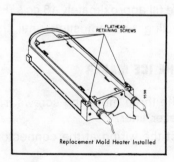

Figure A-29 REPLACEMENT MOLD HEATER

MOTOR REPLACEMENT

1. Remove the ice maker from the cabinet.
2. Remove the mounting plate (see mounting plate removal).
3. Disconnect the two leads to the motor.
4. Remove the motor mounting screws.
5. Install new motor in reverse order, and install ice maker in cabinet.

WATER VALVE SWITCH REPLACEMENT

1. Remove the ice maker from the cabinet.
2. Remove plastic from cover.
3. Remove mounting plate (see mounting plate removal).
4. Disconnect water valve switch leads, (brown and blue leads).
5. Remove the switch mounting screws.
6. Reinstall in reverse order, replace switch insulator.
7. Check water fill, (see text Adjusting The Water Valve Switch).

HOLDING SWITCH REPLACEMENT

1. Proceed as in preceding text.
2. Disconnect switch leads, take note of wire positioning.
3. Remove the two mounting screws.
4. Reinstall switch, (wires must be attached in proper place).
5. Install ice maker in the cabinet, connect wiring.

SHUT-OFF ARM SWITCH REPLACEMENT

1. Proceed as in preceding text.
2. Raise the shut-off arm, remove the three switch leads. Take note of their position.
3. Remove the switch mounting screws.
4. Reinstall in reverse order.

THERMOSTAT REPLACEMENT

1. Remove ice maker from cabinet.
2. Remove the plastic front.
3. Remove mounting plate, note position of shut-off arm spring.
4. Remove thermostat clip mounting screw.
5. Disconnect the wire leads and remove the thermostat.
6. Remove the old alumilastic from the mounting area. Install new thermostat, covering the mounting area with new alumilastic.
7. Connect wire leads, install components in reverse order. Install the ice maker into the cabinet.

MOLD HEATER REPLACEMENT

1. Remove the ice maker as described previously.
2. Remove ice stripper from side of the mold.
3. Remove plastic from cover.
4. Remove mounting plate as in previous text.
5. Remove the thermostat from the mold, disconnect the mold heater wire leads.
6. Remove the mold mounting screws from the support, do not damage the thermostat gasket.
7. Pry the defective heater from the mold.
8. Clean the surface of the old alumilastic.
9. Coat the heater groove with new alumilastic.
10. Install new heater, using the screws provided in heater kit, see Figure A29.
11. Replace parts in reverse order, bond the thermostat with alumilastic. Be sure gasket is in place.
12. Check all wiring, refer to Figure A30, reinstall the ice maker in the cabinet.

SERVICE TIPS AND MAINTENANCE

1. When refrigerator is not in use for a period of time, the ice mold should be drained and dried to prevent oxidization of the mold.
2. The strainer in the water valve will collect sand and dirt and should be cleaned occasionally.
3. When the ice maker is manually run through a cycle, the ejector blade will complete only one revolution. At this time the mold heater will not energize because the thermostat remains open until the proper temperature is reached.

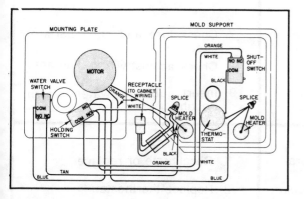

Figure A-30   ICE MAKER WIRING DIAGRAM

WHIRLPOOL DESIGN AUTOMATIC ICE MAKER

FREESTANDING OR UNDERCOUNTER

DIAGNOSIS GUIDE

COMPLAINT	POSSIBLE CAUSE	REMEDY
1. Compressor won't run; no ice in bin	1. Located in cold area.	1. Move to warmer area (above 55°F.).
	2. Power disconnected.	2. Connect power.
	3. Broken wire or loose connection.	3. Locate and repair.
	4. Defective relay.	4. Replace relay.
	5. Service switch in "Off" position.	5. Move operating rod to "On" position.
	6. Bin Thermostat contacts open.	6. Replace bin thermostat.
	7. Defective compressor motor.	7. Replace compressor.
	8. Master Switch in "Clean" position.	8. Push switch to "On" position.
2. Compressor runs; no ice in bin.	1. Water supply shut off.	1. Restore water supply.
	2. Inoperative water valve.	2. Check valve parts for restrictions, binding, or solenoid coil failure.
	3. Evaporator thermostat out of calibration.	3. Recalibrate or replace.
	4. Hot gas solenoid stuck "Open."	4. Repair or replace solenoid.
	5. Inoperative refrigeration system.	5. Repair sealed system.
	6. Excessive use of ice cubes.	6. Advise customer.
	7. Cutter grid circuit open.	7. Check fuse and other parts of circuit.
	8. Incorrect wiring.	8. Check against wiring diagram.
	9. Water inlet tube from valve not inserted in return trough.	9. Insert tube in water return trough.
3. Compressor runs continuously; bin full of ice.	1. Bin thermostat out of calibration.	1. Recalibrate or replace.
	2. Bin thermostat contacts stuck shut.	2. Replace thermostat.
	3. Incorrect wiring.	3. See No. 8 Above.

40

COMPLAINT	POSSIBLE CAUSE	REMEDY
4. Low ice yield.	1. Located in cold area.	1. Move to warmer area above 55° F. (for best results, 70° to 90°).
	2. Water falling on ice cubes.	2. Check water system components and see that they are in proper place.
	3. Bin thermostat out of calibration.	3. Recalibrate or replace.
	4. Evaporator thermostat set to produce too thin or too thick ice cubes.	4. Move adjusting screw to setting to product ½" to ⅝" cube.
	5. Hot gas solenoid stuck partially open.	5. Repair or replace solenoid.
	6. Insufficient refrigeration.	6. Check and repair sealed system.
	7. Not enough water being circulated over evaporator plate.	7. Check for restriction in water lines. Check water pump and motor.
5. Excessive water dripping on ice cubes.	1. Water tank overflowing.	1. Check overflow tube for restrictions. Overflow hose not inserted in liner drain. Incorrect or worn flow washer in water valve.
	2. Water return trough out of position.	2. Install correctly.
	3. Water inlet tube from water valve not inserted in water return trough.	3. Locate tube properly — see pictorial drawing of water system.
	4. Ice jam on cutter grid causing water "to bridge" and drop in bin.	4. Check cutter grid circuit. Check for mineral deposit on evaporator place (See 6).
	5. Water deflector out of position.	5. Install properly.
	6. Water line leak at water valve.	6. Tighten connection to stop leak.
6. Mineral deposit on evaporator plate.	1. High mineral content in water (also see 12).	1. See cleaning instructions.

COMPLAINT	POSSIBLE CAUSE	REMEDY
7. Ice cubes too thin.	1. Evaporator thermostat set for thin cube.	1. Turn thermostat adjusting screw clockwise until cube of desired thickness is obtained.
	2. Not enough water being circulated over evaporator.	2. Check for restriction in water. Check water pump, motor and distributor tube.
	3. Evaporator thermostat bulb on wrong side of shim on evaporator clamp.	3. Shim must be between evaporator bracket and thermostat feeler tube.
	4. Evaporator thermostat out of calibration.	4. Recalibrate or replace.
8. Ice cubes too thick.	1. Evaporator thermostat set at or beyond maximum thickness.	1. Turn thermostat adjusting screw counterclockwise until cube of desired thickness is obtained.
	2. Evaporator thermostat out of calibration.	2. Recalibrate or replace.
	3. In rare cases, due to incorrect positioning of inlet and outlet tubes in header, liquid refrigerant will "spill over" and run down the suction line to compressor. This results in longer operation of the compressor and thicker than normal cubes at times.	3. Remove sufficient liquid refrigerant from system to keep suction line slightly below room temperature during the "Freeze" cycle. This will be approximately 2 ounces.
9. Condenser fan won't run during ice-making cycle.	1. Fan Blade binding on shroud.	1. Adjust shroud to clear fan blade.
	2. Defective motor.	2. Replace motor.
	3. Open circuit in wiring.	3. Locate and repair (see wiring diagram).
	4. Defective evaporator thermostat.	4. Replace thermostat.

SELF CONTAINED ICE MAKER

WHIRLPOOL TYPE

Principally, this ice maker freezes a sheet of ice and cuts the ice into cubes. This is all done automatically.

The cube thickness can be adjusted; the thickness control will accomplish this. At its maximum it will produce 50 lbs. of ice cubes in a twenty-four hour period. This will vary somewhat, depending upon the room temperature where it is installed, and the water temperature as it enters the ice cube maker.

A self contained refrigerator unit, using a freezer plate, is the heart of the ice maker. Water continually flows over and is recirculated over the freezer plate, until a measured thickness of ice slab is produced, Figure SC1 during the freezing cycle of the ice maker. As the thickness of the slab reaches this predetermined thickness, it is sensed by the evaporator thermostat, and the freezing cycle is terminated. At the precise time, the defrost cycle is initiated by energizing the hot gas solenoid valve. Hot gas from the compressor is allowed to flow directly to the freezer plate, thus releasing the slab of ice, which slides down onto the cutting grid which is heated by a low voltage current generated by a transformer. The slab is finally reduced to cubes, while the refrigerator unit once more enters the freezing cycle.

At the start of the defrost period, the remaining water, that had been circulating over the freezer plate and has a high mineral content, is removed through the overflow tube from the water tank reservoir. At this time fresh water enters through the use of a water valve solenoid and the water valve.

The refrigeration unit is charged with Freon 12 and the unit operates much like a conventional refrigerator. A $\frac{1}{8}$ HP rotary type compressor is used to pump the refrigerant vapors from the first discharge tube back into the motor compressor. The vapors are conducted from the compressor discharge into the condenser, where the air circulating over the condenser removes the heat and condenses the vapor back into liquid refrigerant, Figure SC2.

The high pressure refrigerant then passes from the condenser through a capillary tube. A drier is installed just ahead of the capillary tube. The capillary tube, being used as a metering device, controls the flow of refrigerant into self evaporator. From the evaporator the refrigerant flows into an accumulator, where it is drawn off in vapor form by the compressor to start the cycle over again.

As a heat interchanger, the capillary tube uses the lower bottom edge of the freezer plate and is in contact with the freezer plate and soldered to the suction line as it leaves the accumulator. When the hot gas solenoid is energized, the ice slab is released and at that time the condenser fan motor and the water pump is stopped. The compressor continues to pump hot gas directly to the evaporator. The hot gas enters the evaporator at the evaporator and capillary tube junction and raises the temperature to allow the ice slab to release and slide onto the cutter grid; at this time the water valve opens. The evaporator thermostat again returns to the freezing position as the hot gas solenoid is deenergized and the water inlet valve closes. At this time the condenser fan once again is put in motion and the water pump is energized to start the manufacture of a new ice slab.

"THICKNESS" ADJUSTMENT OF ICE SLAB

To increase the thickness of the ice, the evaporator thermostat knob must be turned clockwise. The maximum thickness that it can be adjusted to is $\frac{3}{4}$ of an inch. Turning the knob counter clockwise reduces the thickness. The size of the cubes can be increased only with the purchase of new cutter grids that are available through your distributor. Adjustment of the ice cube thickness should be made only after the ice cube maker has been in operation for at least 24 hours and the storage bin is at least half full of ice cubes.

ACCESS TO COMPONENTS

All functional components can be reached and checked through the front of the ice maker.

BIN DOOR REMOVAL

1. Open part way. By grasping the two sides and pulling up sharply the bin door will release from the hinge pins

INNER BIN DOOR REMOVAL

1. By grasping the door at center bottom and slightly twisting and pulling back, the door will release from the hinge pins.

TOP INSULATED PANEL REMOVAL

1. Lift the flat sheet rubber gasket at the top of the door opening.
2. Remove the two screws.
3. The panel will now drop straight down off of the pins holding in the top of the panel.

PICTORIAL VIEW OF WATER SYSTEM

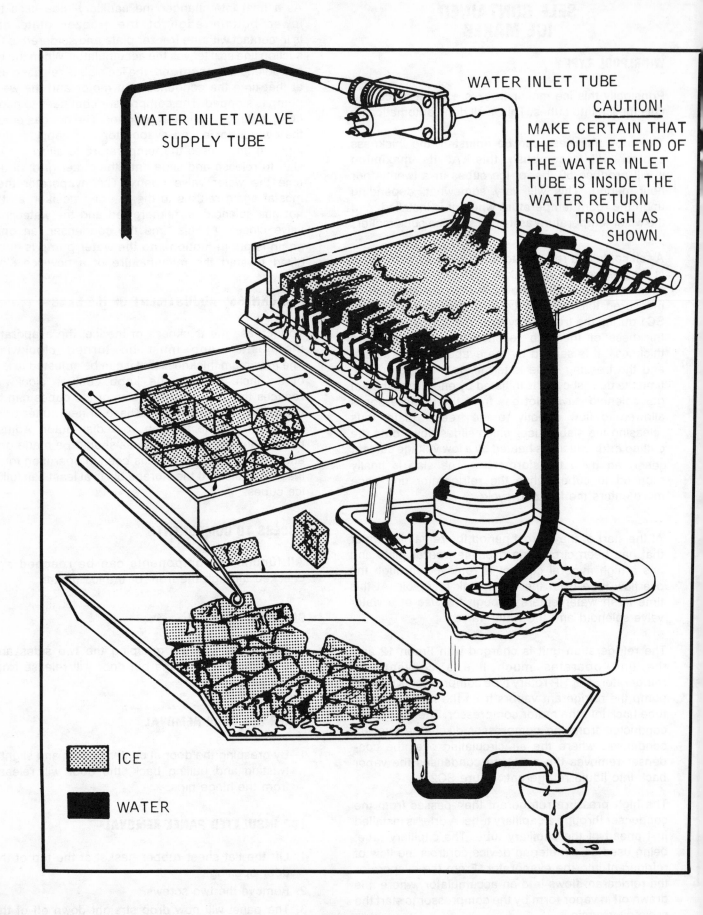

WATER INLET VALVE
SUPPLY TUBE

WATER INLET TUBE

CAUTION!
MAKE CERTAIN THAT
THE OUTLET END OF
THE WATER INLET
TUBE IS INSIDE THE
WATER RETURN
TROUGH AS
SHOWN.

ICE

WATER

Figure SC-1

B. PICTORIAL VIEW OF REFRIGERATION SYSTEM

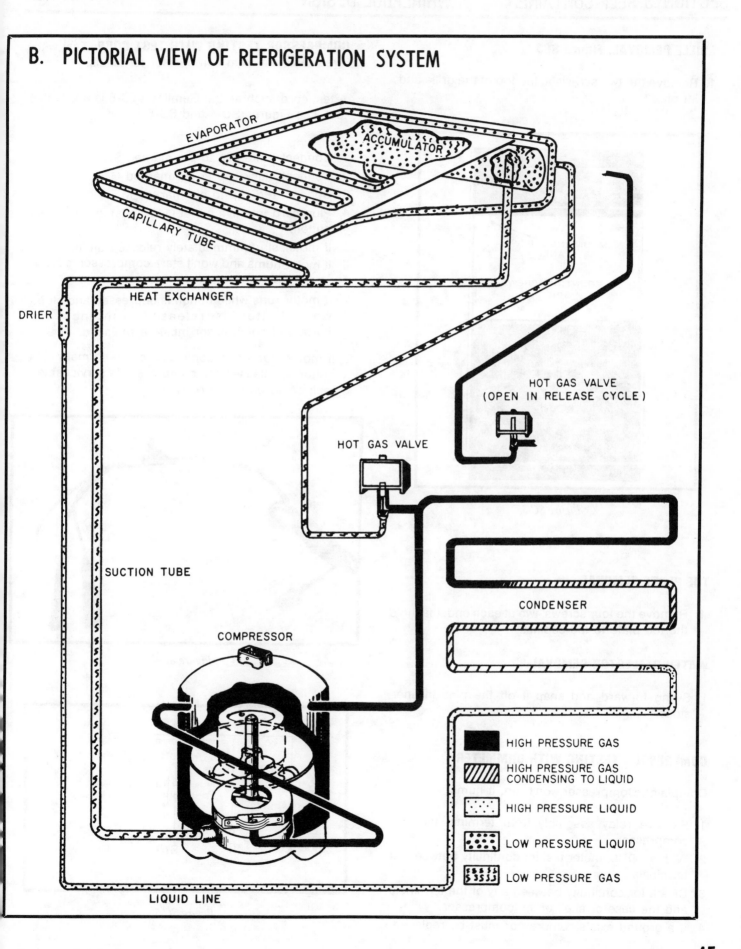

Figure SC-2

GRILLE REMOVAL, Figure SC3

1. Remove the two screws at the top of the grille and lift off.

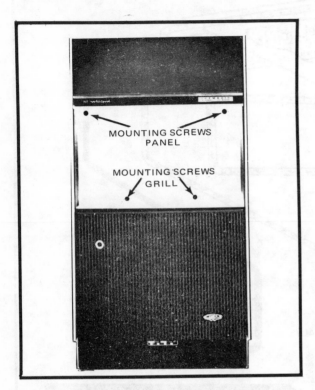

Figure SC-3

TOE PLATE REMOVAL

1. Remove the four screws, two at each end, that hold the toe plate to the cabinet.

WATER DEFLECTOR REMOVAL

1. Swing forward and snap it off the pins in liner sides.

COMPRESSOR TESTING WITH OHMMETER

Complaint, Compressor won t run, it hums.

1. Remove relay assembly from terminal pins on compressor.
2. With an OHM meter, test for continuity between all terminals.
3. Check for continuity between any of the terminals and the case or shell of the compressor.
4. If a ground exists, compressor must be replaced.

COMPRESSOR TESTING WITH TEST CORD. Test with wires removed from compressor terminals.

A test cord such as the Gemline #TC-6 is a valuable tool, see Figures SC-4 and SC-5.

1. Connect white clip to pin marked S.
2. Connect black clip to pin marked G.
3. Connect red clip to pin marked M.
4. Plug cord into live outlet (115 Volts) and depress button.
5. If motor starts immediately release button.
6. If motor hums and won't start, compressor is stuck and should be replaced.
7. If motor runs while button is depressed and stops when button is released, windings are burnt and compressor must be replaced.
8. If motor starts and continues to run normally after button is released the compressor is serviceable, fault could be in the relay.

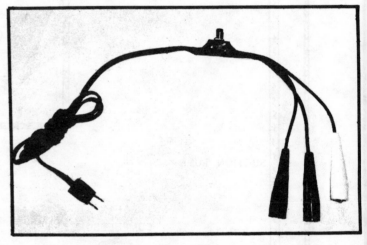

Figure SC-4

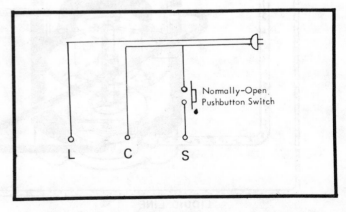

Figure SC-5

RELAY—Complaint, Compressor hums, won't run (Figure SC6)

The function of the relay is to separate the starting winding from the circuit after the motor on the compressor starts. The relay is a magnetic current type relay. To test the relay follow this procedure:

1. Disconnect power source.
2. Remove the wire fastened under terminal C on relay and connect under the same screw terminal with M.
3. Remove wire from terminal S.
4. Have an assistant reconnect power source.
5. Momentarily touch wire removed from S, to terminal M.
   (Above with caution).
6. Remove the S wire from terminal M, if and when the compressor starts.
7. If the compressor starts and runs after the S wire is removed from terminal M, it is indicative that the relay is at fault and should be replaced.
8. Replace relay with proper rated component.

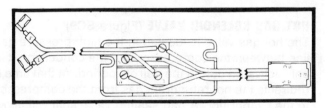

Figure SC-6

EVAPORATOR THERMOSTAT (Figure SC7-SC7A)

The evaporator thermostat has a constant cut-in 38° Plus or minus 2° F. It will cut-out at 10°, to minus 3° F. The 54" capillary is fastened to the underside of the evaporator. In adjusting the control it actually adjusts the ice thickness, as the control allows the unit to run that much longer to attain a thicker slab of ice. When the control senses the evaporator has reached the predetermined temperature, it deenergizes the fan and water pump, at the same instant it energizes the hot gas solenoid and the fill valve solenoid.

TESTING THE EVAPORATOR THERMOSTAT

1. Disconnect the power source.
2. Allow evaporator to reach a higher temperature than 38° or disconnect the control capillary from the underside of the evaporator.

3. Disconnect wires to the control.
4. Continuity should exist between terminals 1 and 2 on the control.
5. If evaporator temperature is allowed to cool to 10° plus or to minus 3°, there should be no continuity between terminals 1 and 2 on the evaporator thermostat.
   IMPORTANT: By disconnecting the water pump and not permitting water to circulate over the evaporator plate, the control will cycle in a few moments. This is the quickest way to observe the action of the thermostat. During this period the solenoid valves will not activate and the condensor fan will continue to run.

6. If the thermostat does not act within the prescribed limits, it would be prudent to replace, rather than try to adjust.

Figure SC-7

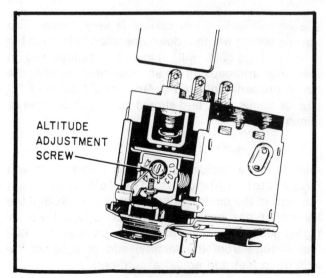

Figure SC-7A

EVAPORATOR THERMOSTAT REPLACEMENT

1. Disconnect the power source.
2. Remove control from the terminal box.
3. Remove the control wires.
4. Remove the cutter grid and water trough.
5. Remove tape securing capillary to the hot gas defrost restrictor tube.
6. Remove clamp from underside of the evaporator plate.
7. Remove the clips under the gasket from the left hand side of the liner.
8. Bend the liner flange forward and remove the capillary tube and the control attached.
9. When installing a new control, the capillary tube MUST be taped to the hot gas restrictor tube, with the rubber shim stock between the clamp and the capillary tube.

BIN THERMOSTAT (Figure SC7A)

The sensing tube and bulb of this thermostat is located in a well on the left side of the liner. When the ice in the bin comes in contact, the thermostat opens and operation of all components, with the exception of the cutter grid, is shut down.

The adjustment of this control is very critical. Too warm a setting will shut down operation before the bin is full, and too cold will allow ice to buildup and hit the cutter grid causing an abundance of water in the bin. A properly adjusted control should cut in at 42° plus or minus 1° F., and should cut out at 36° plus or minus 1° F.

Altitude has a decided effect upon both the bin and Evaporator thermostat. The following chart will help in the control adjustment (Figure SC8). If the ice maker has been recently moved to a lower or higher sea level, (the control comes adjusted for 2000 feet altitude from the factory) add or subtract the altitude in feet and adjust accordingly.

TESTING THE BIN THERMOSTAT

1. Disconnect the power supply.
2. Remove control from terminal box.
3. Disconnect wires at the control.
4. Above 42°F. continuity should exist between the terminals.
5. Below 36°F. there should be no continuity between the control terminals.

ALTITUDE IN FEET	CLOCKWISE TURN OF ADJUSTMENT SCREW
Bin Thermostat	
2000	3/32
4000	9/32
6000	7/16
8000	19/32
Evaporator Thermostat	
2000	$\frac{1}{4}$
4000	$\frac{1}{2}$
6000	$\frac{3}{4}$
8000	1

Figure SC-8

HOT GAS SOLENOID VALVE (Figure SC9)

The hot gas valve is located in the high pressure gas line between the compressor and the condensor. It is only activated during the defrost period. At that time, it redirects the hot refrigerant gas from the compressor to the evaporator. A bad seating valve seat will allow hot gas to leak into the evaporator during the freezing period causing prolonged freeze period in which ice may not even be made. The solenoid coil, if burnt out, will not allow the ice slab to release. The plunger on the valve seats by gravity and is not spring loaded. The valve must be installed vertical with the coil at the top.

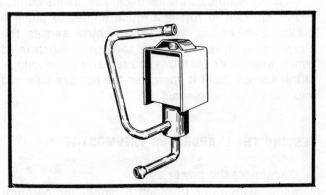

Figure SC-9

TESTING THE HOT GAS SOLENOID VALVE

1. Remove the solenoid wires and check across them for continuity.
2. With a direct test cord, and the power source of the ice maker disconnected, apply the test cord leads to the solenoid coil.
3. Listen for a distinctive click sound as the plunger raises up and is dropped as you disconnect the test cord.
4. The click should be resounding, and not sluggish.
5. With the compressor running, heat should be felt in the hot gas defrost tube when the valve is energized.
6. Disconnect the water pump, deenergize the solenoid, and with the compressor running, the evaporator should frost over and reach a cycling temperature in a few minutes.
7. The complete unit will have to be discharged to replace the solenoid gas valve.

WATER INLET SOLENOID TESTING (Figure SC10)

There should be enough water entering the ice maker during the defrost period to overflow the water pump pan standpipe. During the freeze cycle water should not enter the pump pan. If the valve leaks or allows water to enter during the freeze cycle, it will indicate that the valve needs to be cleaned because of mechanical failure. To determine whether the valve leaks because of mechanical failure or electrical problem, disconnect the water valve solenoid wires (two wires). If water continues to flow, then the problem is mechanical failure of the valve. Lime deposits from the water will build up causing the armature to stick in an open position.

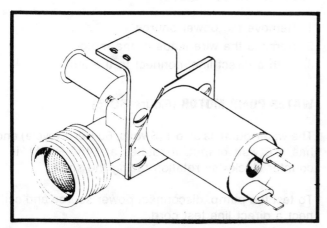

Figure SC-10

To clean the water valve see "Cleaning the Water Valve" on page *13* of this manual. For further information on the water valve read pages *12* thru *13* For disassembly of the water valve, see Figure SC13.

The coil, as used on the water valve, is placed over a brass cylinder that contains a plunger, often referred to as an armature. One end of the plunger is tapered and used as a needle valve. A push spring is placed at the other end of the plunger. When the coil is energized the magnetic force lifts the plunger from the seat allowing water to enter the valve through the orifice. Because of the bleeder hole in the diaphragm, the diaphragm is lifted from the seat and hangs in limbo, allowing a full flow of water to enter the tub, Figure SC11.

If a condition exists that water enters the tub in other cycles than intended, there is an indication that the valve is mechanically stuck open and must be disassembled and cleaned.

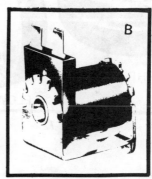

Figure SC-11

CLEANING THE WATER VALVE

NOTE: A good cleaning agent to use on the internal parts of the valve is a 50% solution of soldering acid (can be purchased at a local hardware store) and water. Mix solution in a glass bowl.

1. Remove brass cylinder and diaphragm.

2. Inspect valve seat for any nicks or worn spots.

3. Remove spring and plunger from cylinder.

4. A valve kit can be purchased. The kit contains a cylinder, plunger, spring and a diaphragm.

5. If a new kit is not readily available, dip the above parts in the solution as outlined above in NOTE. *When parts are clean of lime deposits, remove and wash with clean water.*

6. Reassemble to valve. Replace solenoid coil and install on machine.

NOTE: The diaphragm has an orifice hole and a bleeder hole. Figure SC12.

CAUTION: Avoid getting acid on your hands.

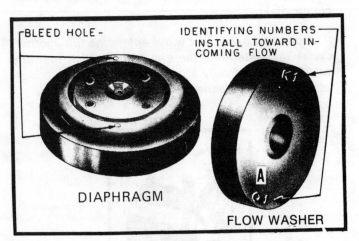

Figure SC-12

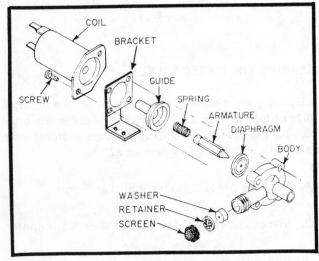

Figure SC-13

Figure SC-14

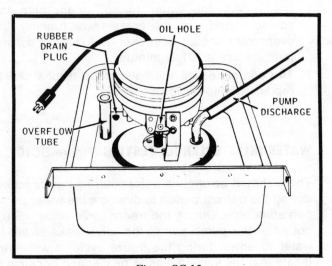

Figure SC-15

FAN MOTOR (Figure SC14)

The fan motor is a 115volt, 60 Hertz (cycle) motor. It has a 1 watt output and turns at 950 R.P.M. in a counter clockwise rotation facing the shaft end.

TESTING THE FAN MOTOR

1. Remove the power source.
2. Remove the wire leads of the motor.
3. With a direct line, connect to the motor.

WATER PUMP MOTOR (Figure SC15)

The water pump is also 115 volts, 60 Hertz (cycle) and has a 6 watt output. It rotates at 1450 R.P.M. in a counter clockwise rotation facing the shaft.

To test the pump, disconnect power source and connect a direct line test cord.

WATER PUMP REPLACEMENT

1. Remove bin door and front insulated panel.
2. Remove inner bin door.
3. Remove the two screws securing the water tank bracket to the liner.
4. Disconnect discharge hose.
5. Disconnect the electrical connections.
6. Remove pump and tank assembly from storage bin.
7. Take out the four screws that secure the motor to the mounting bracket.
8. Remove the motor from the drive clip on the pump impeller.

The motor or the pump can be replaced separately, See Machine Component Parts. After installation of the new component, the impeller should spin freely. The pump motor can be oiled by removing the plug from the oiling port, use one or two drops of SAE 10 motor oil.

MASTER SWITCH

The master switch is a double pole, double throw switch, rated at 15 amps at 250 volts A.C. It is a manually operated switch.

MASTER SWITCH TESTING

1. Remove wires from switch connectors.
2. Place switch in "On" position.
3. Continuity should exist between 1 and 2 terminals and between 3 and 5 terminals.
4. Place switch in "Clean" position.
5. Continuity should exist between 2 and 3 terminals ONLY. Replace if necessary.

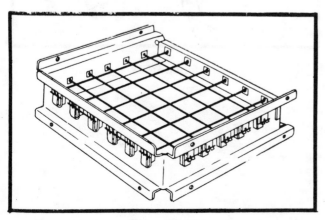

Figure SC-16

CUTTER GRID (SC16)

The cutter grid assembly is so arranged, that the ice slab is cut into square cubes. The wires cross over each other without touching. The transformer current passes through the nichrome resistance wire, which causes the wires to heat and cut through the ice slab.

CUTTER GRID TESTING

1. Carefully examine the grid for broken wires.
2. Carefully examine the connecting pins for corrosion.
3. Look for cracked or broken insulators in the frame.
4. Check for low voltage of the supply line, check voltage at grids.
5. Check fuse.
6. Examine for loose tension of the spring clips.

CUTTER GRID COMPONENTS REPLACEMENT

The wires, spring clips, buss bars and insulators can be replaced separately. If the frame and insulators are broken and a few of the grid wires need replacement, it would be prudent to replace the entire grid. Figure SC17.

1. With the aid of a "C" clamp, compress one of the spring clips to relieve the tension. See Figure SC18.
2. Using a pair of pliers, compress the adjacent spring clip and remove the buss bar.
3. The insulators, clips or wires can now be removed and replaced.

CUTTER GRID TRANSFORMER

The transformer uses line voltage, 115 volts in the primary circuit, the secondary is reduced to 8.5 volts.

CUTTER GRID TRANSFORMER TESTING

Remove the transformer from the circuit.

1. With a test cord, connect directly to the primary wiring and connect to a power source.
2. Using a volt meter, test the secondary, read-out should be 8.5 volts.
3. If transformer fails this test, replacement is necessary.

FUSE

The fuse is located behind the bin door and grille, remove and check for continuity. To remove the fuse, push up and turn, it can be replaced in the same manner. Also check fuse holder for continuity.

CUBE CUTTER GRID

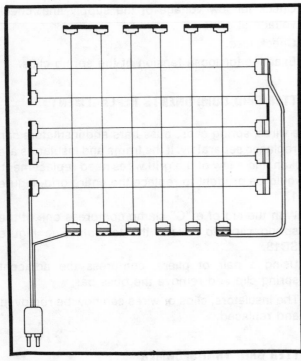

Figure SC-17

THE SEALED REFRIGERATION SYSTEM

The refrigeration system is completely sealed and to effect repairs, the system must be entered. Failures of the sealed system will require tapping into the lines for gauge readings using a line tap valve, or cutting the tubings to make repairs. Correctly diagnosing the problem is the first and most important step before any attempt is made to repair the system. If any of the following occur, it will be necessary to discharge and repair the system.

LEAKS — If it is apparent that the refrigerant has leaked out, a line tap valve should be installed and the system should be partially charged and tested for leaks with a good leak detector.

COMPRESSOR — Failure of the compressor either mechanically or electrically.

RESTRICTIONS — Partial or complete restriction of the capillary tube or dryer.

Master Publications #7552 and 7551 which deals with the refrigeration system is available through your supplier. These Repair-Masters cover the diagnoses, repairs and the principles of refrigeration. It teaches and instructs brazing and welding procedures that are so necessary and important to the repairs of sealed systems.

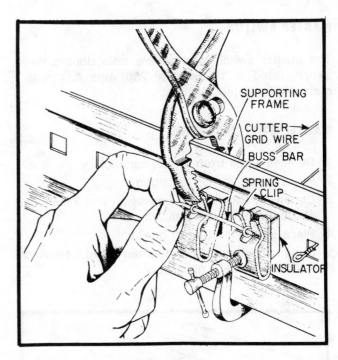

Figure SC-18

WATER CONTROL COMPONENTS

WATER INLET VALVE

See "Water inlet solenoid" under Electrical Components

SUPPLY LINE TO WATER INLET VALVE

1. This is a copper tube that connects from the compressor compartment to the water valve. To replace this tube, remove the clips from under the gasket along left edge of liner.
2. Bend liner flange backwards, this will reveal the tubing.

OUTLET TUBE FROM WATER INLET VALVE

This is a plastic tube connecting the valve to the water return trough.

WATER TROUGH

This is a plastic trough located between the evaporator and the cutting grid, see Figure SC19.

To replace the trough, first remove the cutter grid. Disengage slots, located in each end of the water trough, from the mounting tabs.

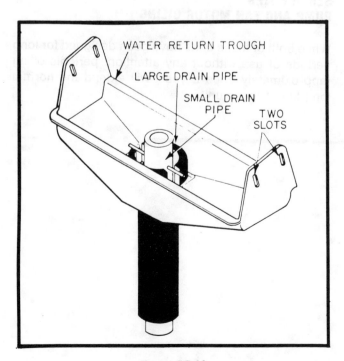

Figure SC-19

WATER TANK

See "Water Pump Replacement"

WATER DISTRIBUTOR

See Figure SC20. This is a ⅝ tube with holes and rubber caps at each end. Its purpose is to distribute the water evenly over the surface of the evaporator plate, to produce an ice slab.

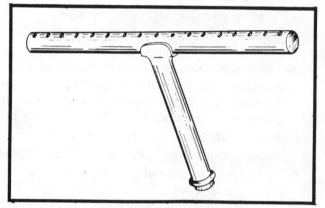

Figure SC-20

WATER DISTRIBUTOR REPLACEMENT

1. Remove the bin door and upper front panel.
2. Remove the cutter grid and water deflector.
3. Disconnect the water tube.
4. Remove distributor, replace in reverse manner.

CABINET COMPONENTS
DOOR CATCH

This is a spring-loaded roller located on the bottom edge of the insulated front panel.

DOOR CATCH (ADJUSTMENT)

1. Remove insulated front panel.
2. Loosen two slotted head screws on inside of insulated front panel directly behind roller catch.
3. Move roller in or out to adjust.

DOOR CATCH (REPLACEMENT)

1. Remove insulated front panel.
2. Remove screws holding door catch, plastic liner and stainless steel deflector to front panel.
3. Slide roller assembly from panel.

LINER (REPLACEMENT)

1. Remove insulated front panel.
2. Remove screws holding door catch, plastic liner and stainless steel deflector to front panel.
3. Slide roller assembly from panel.

LINER (REPLACEMENT)

1. Remove bin door, inner bin door, top insulated front panel and grille.
2. Remove electrical terminal box cover.
3. Remove three screws holding electrical box to cabinet and let it hang.
4. Remove all components from interior of liner.
5. Remove drain from liner bottom by prying metal retainer up and pushing soft molded rubber drain down into insulation.
6. Remove cabinet panel behind electrical box.
7. Remove clips under gaskets from edge of liner.
8. Pull liner from cabinet.

NOTE: Foam insulation must be in place behind door slides to prevent fiberglas insulation from tearing.

GASKET (REPLACEMENT)

1. Remove bin door, top insulated panel and grille.

2. Remove clips from under gasket holding edge of liner.

3. Bend liner edges forward and remove gasket.

WATER TREATMENT

During the freezing process, as water passes over the freezing plate, the impurities in the water have a tendency to be rejected and the plate will freeze only to pure water.

However, the more dissolved solids in the water, the more troublesome the freezing operation will be. Bicarbonates in the water are the most troublesome of the impurities.

These impurities will cause scaling on the freezer plate, clogging of the water distributor head, water inlet valve and other parts in the water system.

If the concentration of impurities is high, cloudy cubes or mushy ice may be the result.

Parts of the ice maker that are in contact with the water or ice, may corrode if the water is high in acidity.

SERVICE TIPS
PUMP AND FAN MOTOR OILING

While both of these units have been designed for long periods of use without any attention, periodic oiling (approximately once a year) will extend the normal long life of these units.

GENERAL ELECTRIC DESIGN

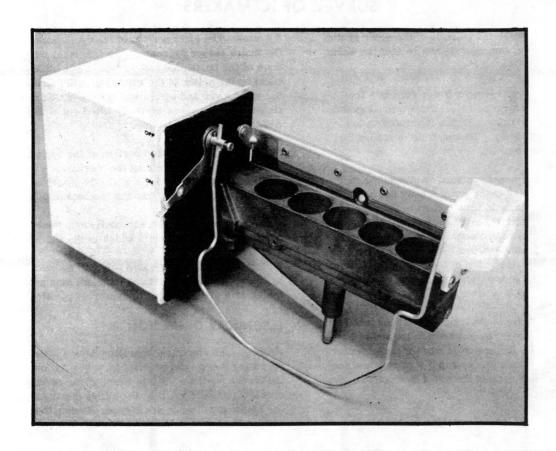

Used by

HOTPOINT
NORGE (Early "60" Models)

The following pages illustrate the ice makers used
on the corresponding refrigerator models.

SURVEY OF ICEMAKERS

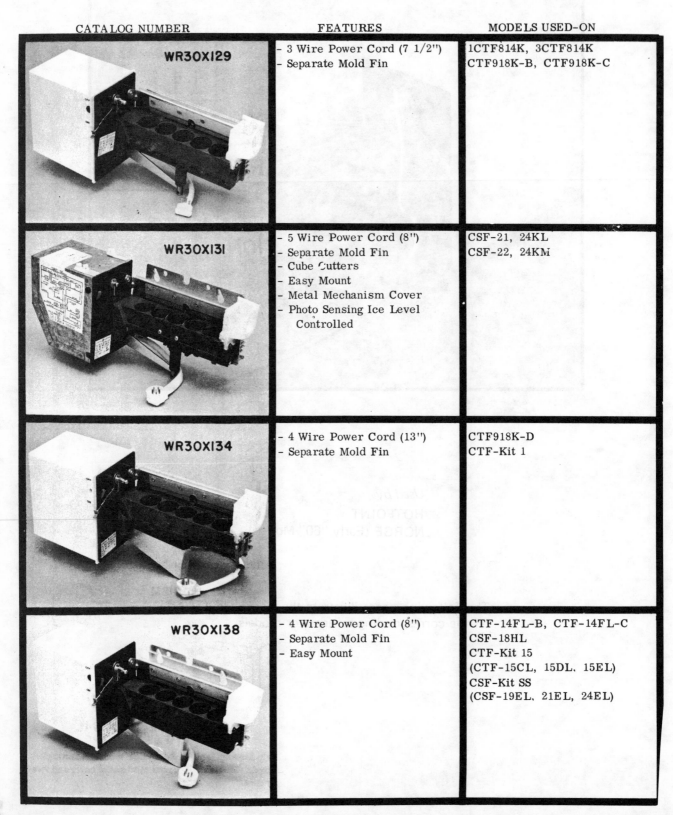

CATALOG NUMBER	FEATURES	MODELS USED-ON
WR30X129	- 3 Wire Power Cord (7 1/2") - Separate Mold Fin	1CTF814K, 3CTF814K CTF918K-B, CTF918K-C
WR30X131	- 5 Wire Power Cord (8") - Separate Mold Fin - Cube Cutters - Easy Mount - Metal Mechanism Cover - Photo Sensing Ice Level Controlled	CSF-21, 24KL CSF-22, 24KM
WR30X134	- 4 Wire Power Cord (13") - Separate Mold Fin	CTF918K-D CTF-Kit 1
WR30X138	- 4 Wire Power Cord (8") - Separate Mold Fin - Easy Mount	CTF-14FL-B, CTF-14FL-C CSF-18HL CTF-Kit 15 (CTF-15CL, 15DL, 15EL) CSF-Kit SS (CSF-19EL, 21EL, 24EL)

Figure A

SURVEY OF ICEMAKERS (Continued)

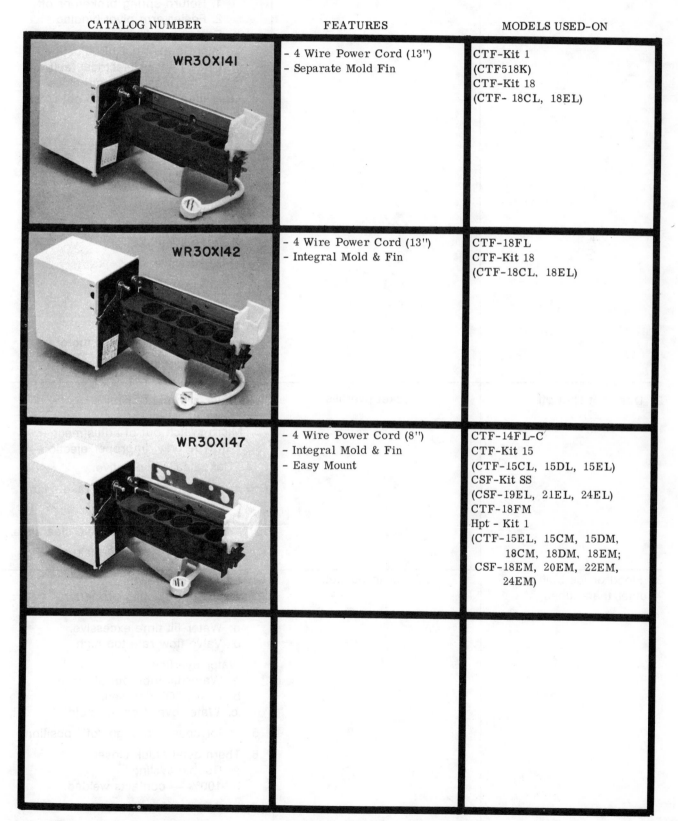

CATALOG NUMBER	FEATURES	MODELS USED-ON
WR30X141	- 4 Wire Power Cord (13") - Separate Mold Fin	CTF-Kit 1 (CTF518K) CTF-Kit 18 (CTF- 18CL, 18EL)
WR30X142	- 4 Wire Power Cord (13") - Integral Mold & Fin	CTF-18FL CTF-Kit 18 (CTF-18CL, 18EL)

WR30X147	- 4 Wire Power Cord (8")
	 - Integral Mold & Fin
	 - Easy Mount
		CTF-14FL-C
		 CTF-Kit 15
		 (CTF-15CL, 15DL, 15EL)
		 CSF-Kit SS
		 (CSF-19EL, 21EL, 24EL)
		 CTF-18FM
		 Hpt - Kit 1
		 (CTF-15EL, 15CM, 15DM,
		      18CM, 18DM, 18EM;
		   CSF-18EM, 20EM, 22EM,
		      24EM)

Figure B

COMPLAINT	SYMPTOM	CAUSE

1. Does not make ice	No ice	1. No power to ice maker
		2. Manual switch off
		3. Not plumbed in, water off or restricted
		4. Feeler arm switch open
		   a. Feeler arm not in down position
		     1. Return spring broken or off
		     2. Feeler arm pivot binding
		   b. Feeler arm bent
		   c. Micro switch out of adjustment,
		     blade bent or electrical fault
		5. Safety thermostat will not close
		6. Operating thermostat will not close
		   a. Out of calibration
		   b. Inoperative
		   c. Freezer not cold enough
		7. Leaf switch inoperative.
		   a. Improperly adjusted.
		   b. Electrical fault.
		8. Water valve not actuated.
		   a. Open solenoid.
		   b. Stuck valve.
		   c. Wiring fault.
		   c. Wiring fault.
		9. General — electrical.
		   a. Inoperative heater or motor.
		   b. Open wiring.

2. Does not shut off	Bucket overfills	1. Feeler arm bent or missing.
		2. Feeler arm blocked.
		3. Micro switch out of adjustment.
		4. TCF type only: Improper ejection.
		   a. Rake snap inoperative.
		   b. Cubes off mold before rake snap.
		     1. Bad heater.
		     2. Excessive pad to mold clearance.
		     3. Forward tilt of ice maker.

3. Flood or ice built up—other than cubes	Water in bucket	1. Leaky water valve.
		2. Cubes not ejecting properly.
		3. Over-fill.
		   a. Water fill time excessive.
		   b. Valve flow rate too high.
		4. Water overflow.
		   a. Water fill tube out of place.
		   b. Leaky "O" ring seal.
		   c. Water over front of mold.
		5. Motor coasts through "off" position.
		6. Thermostat stuck closed.
		   a. Double cycling.
		   b. 100% — contacts welded.
		7. Leaf switch fails closed.
		8. General wiring faults.

COMPLAINT	SYMPTOM	CAUSE
4. Does not eject cubes.	Water or ice in freezer or bucket	1. Wiring 2. Defective solenoid valve 3. Bent leaf switch 4. No water hookup 5. Plugged valve 6. Ice in fill tube

5. Cycles but does not fill with water.	No ice	1. Inadequate rake sweep, insufficient travel
		   a. Wear
		   b. Plastic extension defective
		   c. Misc. mechanical faults

6. Oversize cubes	Mushroom head cubes, usually in clusters of 3-5	1. Water fill time excessive
		2. Leaky water valve
		3. Multiple cycle
		   a. Thermostat does not open
		   b. Motor coasts through off intermittently
		4. Water valve flow rate excessive
		5. Improper cube ejection intermittently

7. Small cubes (under-fill)		1. Water fill time short 2. Low water pressure 3. Plugged water valve 4. Restricted or leaky supply line 5. Water valve flow rate low

8. Unacceptable cube condition	Odor, lumping, frost, sublimation, color	1. Odor
		   a. Lengthy storage
		   b. Unusual contamination
		      1. Odor transfer
		      2. Water supply
		2. Lumping (extended time)
		   a. Wet cubes
		      1. Liquid centers
		      2. Too much melt on eject
		      3. Long bucket removal or door opening
		      4. Refrigeration off
		3. Sublimation, frost, etc.
		   a. Long storage
		4. Color
		   a. Water supply
		   b. Long inoperation periods

EARLY HOTPOINT AND GENERAL ELECTRIC ICE MAKERS

There are two types of ice makers used on early models of Hotpoint and General Electric refrigerators. Although the ice molds and the mechanism looked alike, they had their differences. Shown below are the two types. Figure 1 illustrates the early identification label, while Figure 2 is a later model version of the same ice maker. To find the location of the identification label, see Figure 3. Take note of the difference for quick identification between Figure 4 and Figure 5.

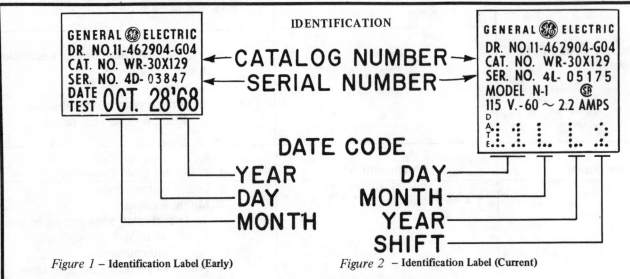

Figure 1 – Identification Label (Early)

Figure 2 – Identification Label (Current)

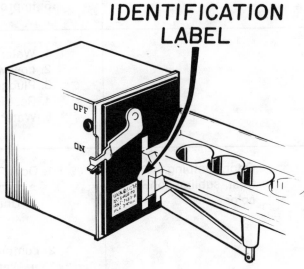

Figure 3 - Location of Identification Label

TYPES

Two icemaker types are currently used in Hotpoint refrigerators for specific application requirements.

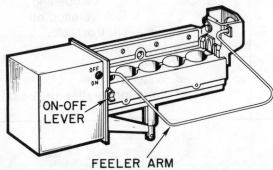

Figure 4 – Feeler Arm - Used on CTF and most CSF Models - identified by feeler arm and manual "on-off" lever.

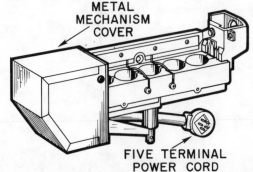

Figure 5 – Photo Sensing Type - Used on CSF models featuring Exterior Ice Service - identified by metal mechanism cover and five terminal cord.

INSTALLATION

The installation of this ice maker and hook up is basically the same as in previous chapters of this Repair-Master. It is important that the local plumbing codes and requirements be followed. Always allow extra tubing as shown in Figure H6, so the refrigerator can be moved for servicing. Most ice makers have an "on-off" lever located on the ice maker housing. The ice maker is "on" when the lever is down. Exterior Ice Service models, with photo sensing ice makers, have an "on-off" knob at the photo switch on the left breaker strip, as shown in Figure H7.

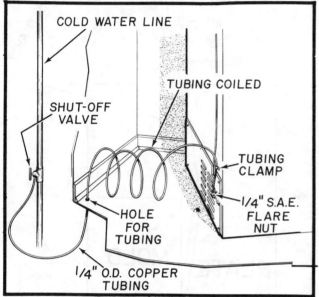

Figure H6
TYPICAL PLUMBING INSTALLATION

If trouble is caused by sand and impurities getting into the water valve, this problem can be alleviated by installing a filter in the water line such as the Gemline Part No. IWF-101. This filter incorporates a replaceable charcoal cartridge which offers an effective control, Figure H8.

Figure H7
EXTERIOR ICE SERVICE CONTROL

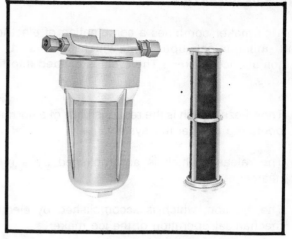

Figure H8

RANGE OF OPERATION, FEELER ARM, FEELER ARM SWITCH

PLATE ASSEMBLY

Feeler Arm Switch

1. Place ice maker in normal reset position.

2. Replace feeler arm with a pointer, and position gauge on the plate to check switch operation. Take note of the points of operation of feeler arm switch-off and on. Adjust position of switch to close contacts as feeler arm pivot passes between 20° and 14°. Make this adjustment in one direction only — returning to normal rest, see Figure H9. Tighten lock screw to 8 inch pounds. Do not over tighten.

PLATE — Mold Water Seal

1. Use WR97X140 sealer to seal between mold and plate on motor side of plate.

MOLD ASSEMBLY

Mold Heater

1. When installing a mold heater, fill the entire length of groove in mold with sufficient WR97X161 mastic to prevent voids and air space. Install heater and remove excess material carefully.

2. Turn heater on for about two minutes to partially cure mastic.

COMPONENT DESCRIPTION AND CYCLE OF OPERATION

The ice maker combines a combination of electrical and mechanical operations which makes it an "automatic ice maker". The cycle is divided into five periods.

1. The freeze, which is the responsibility of a normally operating refrigerator system.

2. The release, which is accomplished by a mold heater.

3. The ejection, which is accomplished by electro-mechanical operation of the ice maker.

4. The sweep, this is the responsibility of the ejector pad, by electro-mechanical means.

5. The water fill, controlled electrically by the use of a solenoid type of water valve.

MECHANICAL OPERATION

1. The first step in the mechanical operation begins with the mold filled with water.

2. In freezing, the heat is removed from the water and the freezing period of the cycle takes place. When the temperature reaches approximately 16°, the operating thermostat closes, and the release period is initiated.

3. Release of the ice pieces starts with the heater and motor being energized. The motor begins rotation, but almost immediately is stalled by the frozen ice until the heater melts the ice pieces free from the mold.

4. The ejection period of the cycle takes place when the motor torque overcomes the frictional resistance of the frozen ice. The ice pieces are lifted out of the mold by the action of the motor. This motor action is exerted through the eject lever and the ejection pad. Ejection continues until all the pieces are lifted from the mold. The motor may stall again and again if some of the ice pieces are slow in releasing.

5. The sweep period of the cycle takes place after all of the ice pieces are ejected to the top of the mold by the ejection pad. The rake sweeps the ice pieces from the top of the mold into the ice bucket.

6. Water fill takes place after the sweep period. The water valve is energized by the rotation of the cam as the eject arm and feeler arm are returning to their rest positions.

RANGE OF OPERATION OF FEELER ARM AND FEELER ARM SWITCH OF ICEMAKER

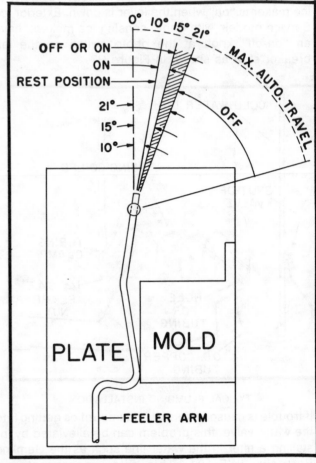

Figure H9

ELECTRICAL OPERATION

1. Freeze: The electrical operation begins when the operating thermostat reaches its preset setting during the freeze period. At this time, (under normal conditions) both the safety thermostat and the feeler arm switch are closed. During the freeze period the leaf switches and the cam are in Position 1, Figure H10.

2. Release takes place with the closing of the operating thermostat. A series circuit is completed through the safety thermostat, the feeler arm switch and the operating thermostat, to energize the motor and the heater. This starts the release period of the cycle.

1. Ejection takes place as the cam rotates at the beginning of the ejection period. Switch number 1 is closed, thus completing a circuit around the operating thermostat. This is the holding circuit of the leaf switch. Position 2, Figure H11. When switch number 1 is closed, the period will continue to its end and the motor and heater will be energized, regardless of the position of the operating thermostat.

2. Under normal operating conditions, the operating thermostat will open in one to one and one half minutes after the beginning of the period. The start of this period was initiated by the closing of the operating thermostat and/or the closing of the feeler arm switch.

3. Water Fill: At about 270° position of the rotating cam, the water fill switch or switch number 2 of the leaf switch, closes for approximately six seconds. Positions 3 and 4, Figures H12 and H13. This allows the required amount of water to fill the mold for the next cycle.

4. Just before the 360° position of the rotating cam, the actuator leaf of the leaf switch assembly falls into the open area of the cam, thus opening switch number 1. By this time, the operating thermostat has opened and there is no longer a circuit to the motor of the ice maker. The motor will coast slightly and then stop. This completes the cycle of operation.

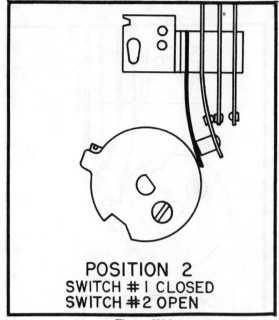

POSITION 2
SWITCH # I CLOSED
SWITCH # 2 OPEN

Figure H11

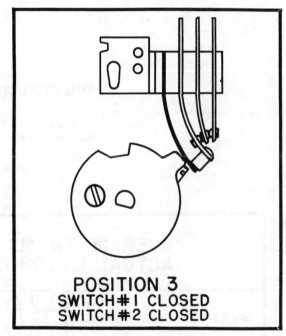

POSITION 3
SWITCH# I CLOSED
SWITCH# 2 CLOSED

Figure H12

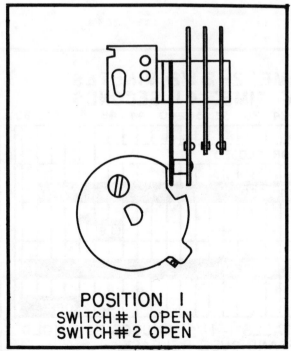

POSITION I
SWITCH# I OPEN
SWITCH# 2 OPEN

Figure H10

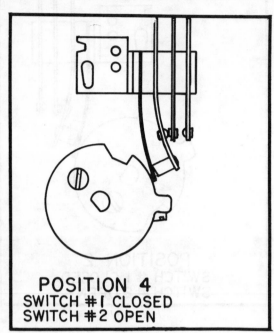

POSITION 4
SWITCH #1 CLOSED
SWITCH #2 OPEN

Figure H13

SCHEMATIC OF ICE MAKER

The schematic of the ice maker is illustrated in Figure H15 and the parts breakdown is shown.

TIME CYCLE CHART

In Figure H14 the average time of each period is illustrated and the positions of the leaf switches are noted.

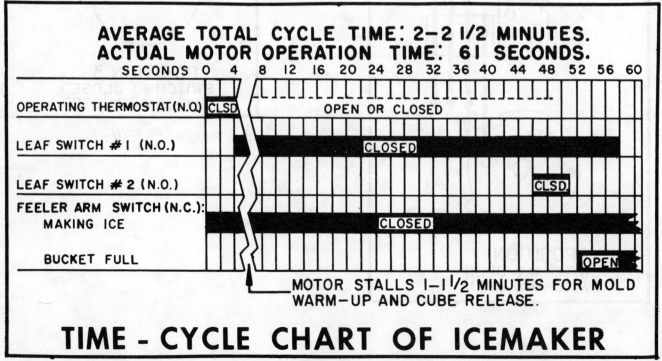

AVERAGE TOTAL CYCLE TIME: 2—2 1/2 MINUTES.
ACTUAL MOTOR OPERATION TIME: 61 SECONDS.

SECONDS 0 4 8 12 16 20 24 28 32 36 40 44 48 52 56 60

OPERATING THERMOSTAT (N.O.) — CLSD — OPEN OR CLOSED

LEAF SWITCH #1 (N.O.) — CLOSED

LEAF SWITCH #2 (N.O.) — CLSD.

FEELER ARM SWITCH (N.C.): MAKING ICE — CLOSED

BUCKET FULL — OPEN

MOTOR STALLS 1—1 1/2 MINUTES FOR MOLD WARM—UP AND CUBE RELEASE.

TIME - CYCLE CHART OF ICEMAKER

Figure H14

ICE MAKER MODEL REFRIGERATORS CTF918K, CTF814K

DATA

Operating Thermostat limits, °F	$26° - 16°$
Safety Thermostat limits, °F	$115° - 50°$
Mold Heater watts	225
Motor rpm	1
watts	26.45
Ejection Cycle minutes per cycle	2-2 1/2
Ice Cube Rate pounds per cycle	.2
pounds per 24 hours (avg.)	4 1/2-5 1/2

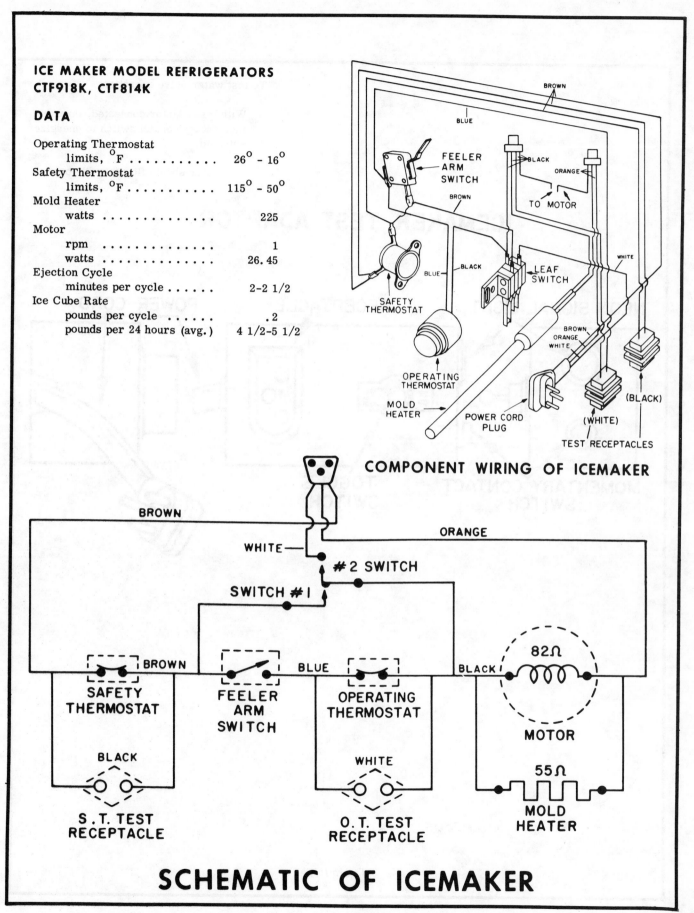

COMPONENT WIRING OF ICEMAKER

SCHEMATIC OF ICEMAKER

Figure H15

To test water valve operation.

With test adaptor connected, operate momentary contact switch to energize solenoid.

Water should flow into fill-cup.

ICEMAKER TEST ADAPTOR

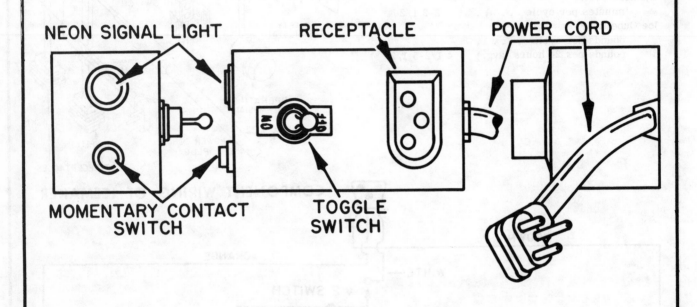

NEON SIGNAL LIGHT   RECEPTACLE   POWER CORD

MOMENTARY CONTACT SWITCH   TOGGLE SWITCH

Figure H19

ELECTRICAL AND OPERATIONAL TEST PROCEDURES FOR ICE MAKER

Tools required: Volt-ohm Meter
Ice Maker Test Adaptor (Robinair #14298)
Ice Maker Schematic
2-Test Receptacle Shorting Tools

The following tests are made with ice bucket removed, but without removing ice maker from refrigerator, and with removing motor cover.

ICEMAKER RECEPTACLE

OPERATING THERMOSTAT TEST RECEPTACLE

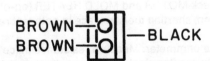

SAFETY THERMOSTAT TEST RECEPTACLE

ELECTRICAL TEST PROCEDURE (with ice maker plug disconnected from receptacle)

1. Check for PROPER VOLTAGE at ice maker receptacle in freezer.

   a. Use voltmeter: Measure voltage between brown lead and *orange* lead sockets.

      (1) Reading should be approximately 115 volts (line voltage)

2. To check OPERATING THERMOSTAT (mold temperature below 13°F., on-off lever in *off* position, ejector pad in rest position)

   a. Use ohmmeter: Test for circuit continuity between *green* lead and *blue* lead sockets of the *white* test receptacle.

      (1) Continuity (0 ohms) indicates operating thermostat is closed.

3. To check SAFETY THERMOSTAT — ice makers with two test receptacles — black and white (mold temperature below 30°F., on-off lever in *off* position, ejector pad in rest position)
   (Note: For ice makers with one test receptacle, see No. 4.1.)

   a. *Use ohmmeter: Test for circuit continuity between the two brown lead sockets of the black test receptacle.*

      (1) Continuity (0 ohms) indicates safety thermostat is closed.

4. To check FEELER ARM SWITCH — ice makers with two test receptacles — black and white (on-off lever in on position, ejector pad in rest position)
   (Note: For ice makers with one test receptacle, see Number 4.1.)

   a. Use ohmmeter: Test for circuit continuity between the *two brown* lead sockets of the *black* test receptacle.

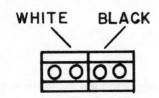

WHITE   BLACK

TEST RECEPTACLES

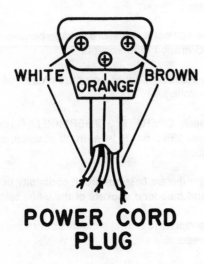

WHITE   ORANGE   BROWN

POWER CORD PLUG

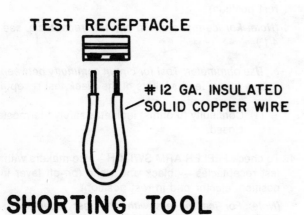

TEST RECEPTACLE

#12 GA. INSULATED
SOLID COPPER WIRE

SHORTING TOOL

(1) If both thermostats are closed, determined in 2 and 3 above, and a circuit is indicated, raise and lower feeler arm.

   (a) Feeler arm switch should operate, opening and closing circuit.

(2) If one or both thermostats are open, place shorting tool in black test receptacle, then check for circuit between terminal for *brown* lead at ice maker plug and either socket of white test receptacle.

   (a) When circuit is indicated, raise and lower feeler arm to operate feeler arm switch, thus opening and closing circuit.

4.1 To check SAFETY THERMOSTAT and FEELER ARM SWITCH — *ice makers with one test receptacle — white* (mold temperature below 30°F., on-off lever in *on* position, ejector pad in rest position).

a. Use ohmmeter: Test for series circuit through feeler arm switch and safety thermostat by checking between terminal for *brown* lead at ice maker plug, and either socket of the white test receptacle.

   (1) Continuity at both terminals of test receptacle indicates safety thermostat, feeler arm switch, and operating thermostat are closed.

   (2) Continuity at only one socket of test receptacle indicates operating thermostat is open.

   (3) When circuit is indicated, raise and lower feeler arm.

   (a) Feeler arm switch should operate, opening and closing circuit.

5. To check MOTOR and MOLD HEATER (on-off lever in *on* position, shorting tools in place in both test receptacles).

a. Use ohmmeter: Measure resistance at ice maker plug between terminal for *orange* lead and terminal for *brown* lead.

   (1) 30-33 ohms indicates both motor and heater are in the circuit (a parallel circuit).

   (a) 80 ohms — motor only

   (b) 55 ohms — heater only

WHITE   BLACK

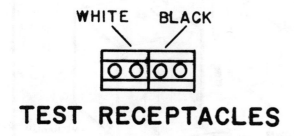

TEST RECEPTACLES

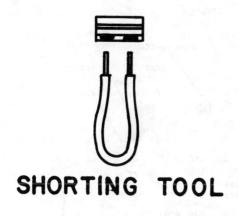

SHORTING TOOL

b. Operational Test Procedure.

1. To test cycle of operation (ice cubes frozen in mold, on-off lever in *on* position, ejector pad in rest position).

   a. With test adaptor connected (test adaptor plugged into ice maker receptacle and ice maker plugged into adaptor), insert shorting tools into test receptacles and turn test adaptor *on*.

      (1) As ice maker operates, observe portions of cycle:

         (a) Release — When test adaptor is first turned on, motor should operate momentarily, then stall; mold should warm slightly and the cubes release.

         (b) Ejection — Motor should operate ejector mechanism raising cubes to top of mold. *REMOVE SHORTING TOOLS DURING EJECTION.*

         (c) Sweep — When cubes reach top of mold, rake mechanism should operate, sweeping cubes from ejector pad into ice bucket.

         (d) Water fill — As ejector pad lowers and feeler arm is near top of its swing, water fill should occur for 6 seconds (indicated by light on test adaptor).

         (e) End of cycle — Operation should stop within 8-10 seconds after water fill is complete.

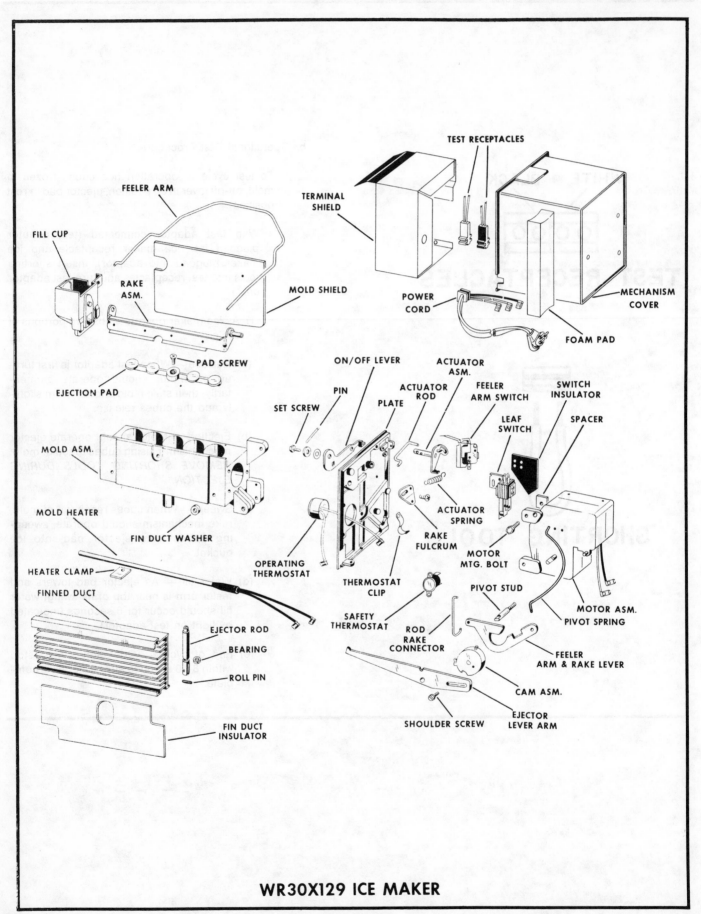

WR30X129 ICE MAKER

Figure H16

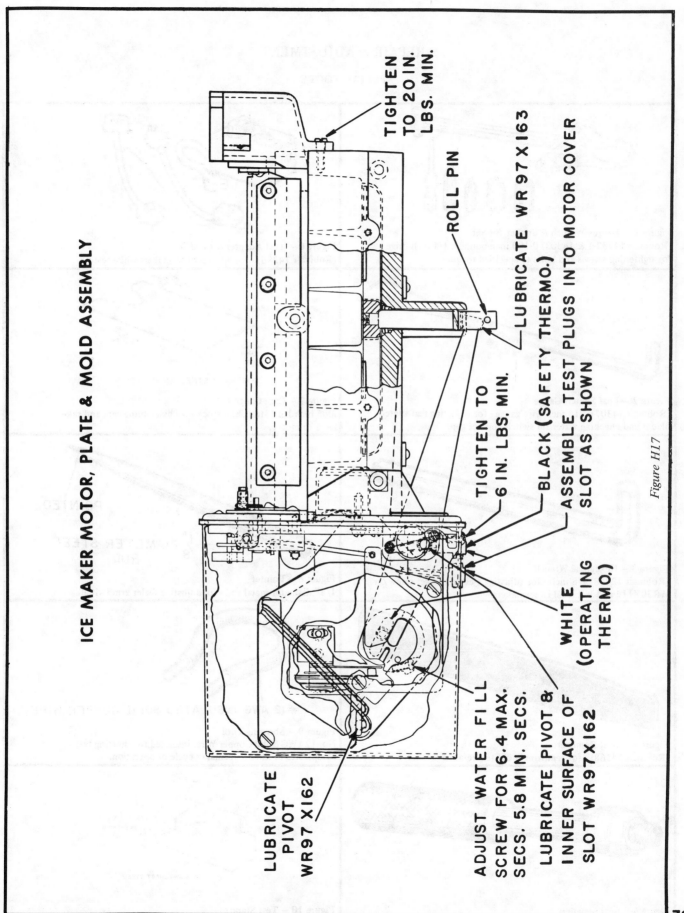

ICE MAKER MOTOR, PLATE & MOLD ASSEMBLY

TIGHTEN TO 20 IN. LBS. MIN.

ROLL PIN

LUBRICATE WR 97 X 163

LUBRICATE WR 97 X 163

ASSEMBLE TEST PLUGS INTO MOTOR COVER SLOT AS SHOWN

BLACK (SAFETY THERMO.)

TIGHTEN TO 6 IN. LBS. MIN.

WHITE (OPERATING THERMO.)

LUBRICATE PIVOT & INNER SURFACE OF SLOT WR97X162

ADJUST WATER FILL SCREW FOR 6.4 MAX. SECS. 5.8 MIN. SECS.

LUBRICATE PIVOT WR97 X162

Figure H17

REPAIR & ADJUSTMENT

SPECIAL TOOLS

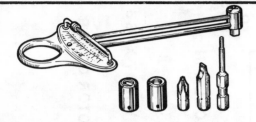

Figure 1 – Torque Wrench & Socket/Bit Set
(Robinair 14232-1 & 14303) 0-100 inch-pounds, 1/4 inch drive - for tightening screws and bolts to specified torques.

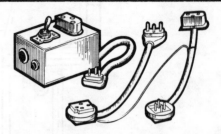

Figure 6 – Test Adapter & Cord Set
(Robinair 14298) - for control of icemakers while testing.

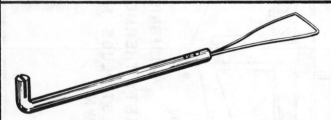

Figure 2 – Leaf Switch Gauge
(Robinair 14307) .016 inch wire gauge - for adjusting leaf switch blades and checking water fill switch contact gap.

Figure 7 – Cam Gauge
.006 inch hard steel shim stock - for assembling cam and rake lever to motor.

.006 HARD STEEL SHIM STOCK

Figure 3 – Hexagonal Wrench
(Robinair 12048) 1/8 inch - for adjusting ejector rod set screw on WR30X112 and WR30X113 icemakers.

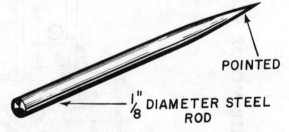

POINTED

$\frac{1}{8}$" DIAMETER STEEL ROD

Figure 8 – Pointer
1/8" diameter metal rod - for adjusting feeler arm switch.

Figure 4 – Seal Remover
(Robinair 14368) - for removing mold seal retainer washer.

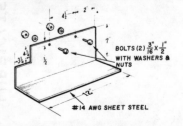

#12 AWG INSULATED SOLID COPPER WIRE

Figure 9 – Shorting Tool
No. 12AWG Solid Copper Wire, insulated for shorting test receptacles to initiate icemaker cycle of operation.

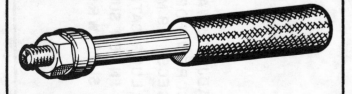

Figure 5 – Seal Installer
(Robinair 14173) - for installing mold seal, bearing and retainer washer.

BOLTS (2) $\frac{3}{16}$" X $\frac{1}{2}$" WITH WASHERS & NUTS

#14 AWG SHEET STEEL

Figure 10 – Test Stand
Metal Bracket - for supporting icemaker in level position while testing.

FRIGIDAIRE DESIGN

ICE-I

SOLID STATE– FLEX TRAY ICE MAKER

ELECTRICAL

Electrical controls are assembled on a printed circuit board which mounts on the front housing. Figure SS-1.

The electrical power supply, the fill tube heater and the water valve are connected to the circuit board through an edge connector.

Two basic electrical systems are integrated on the printed circuit board.

1. Solid State electronic circuitry is provided to sense temperature and thus initiate the harvest cycle. The power supply to this section of the printed circuit board is wired in series with the fill tube heater element which has a resistance of 2200 ohms. Thus the voltage is reduced to approximately 27 volts AC.

2. Line voltage AC circuitry is provided to operate the drive motor and the water valve and to energize the fill tube heater. The power supply to this section of the printed circuit board is directly connected through the edge connector to the AC line.

These two electrical systems are integrated at a single pole, double throw relay which is triggered by the solid state electronic temperature sensing circuitry. The electrical contacts within the relay are in the line voltage AC circuit to the drive motor.

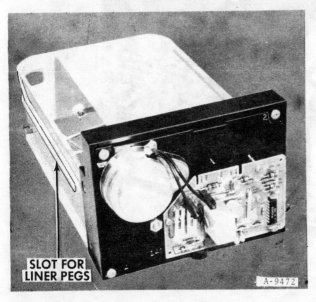

Figure SS-1, The ICE-1 Ice Maker Assembly

SLOT FOR LINER PEGS

A-9472

The temperatuer sensing circuitry utilizes a thermistor which is located within a well at the front of the ice tray. The thermistor senses the temperature of the center cube cavity in the second row from the front

of the tray, Figure SS-2. When the thermistor senses a reduction in temperature to approximately +13° F. at this point, the relay is energized. Electrical contacts within the relay close completing the line voltage circuit to the drive motor. This initiates the harvest cycle.

During the harvest cycle, the operation of the ice maker is controlled through a series of switches in the line voltage circuits to the drive motor and water valve. These switches are operated through push pins which follow cams molded into the front surface of the cam gear.

ELECTRICAL CONTACT CIRCUIT AND SCHEMATIC

NEUTRAL

MOTOR   WATER VALVE

118 VAC

ICE LEVEL

CLOSES TO #1 AT +13° ±3°F
CLOSES TO #2 AT +34° ±2°F

RELAY   +   1 COLD   −   2 WARM

TO DIGITAL TIMER   CLEAR TIMER

STOP   DELAY   HOLD   FILL

LINE

A-9567

SCHEMATIC SHOWN IN TRAY LEVEL AT HOME POSITION

Figure SS-4, The ICE-1 Electrical Schematic

An additional "ice level" switch, Figure SS-3, is provided in the drive motor circuit. This switch has two functions.

1. To stop the operation of the ice maker prior to the ejection of cubes from the tray, should the ice container be removed.

2. To stop the operation of the ice maker when the ice container is full.

This switch is activated by a "push pin" which follows an ice level cam.

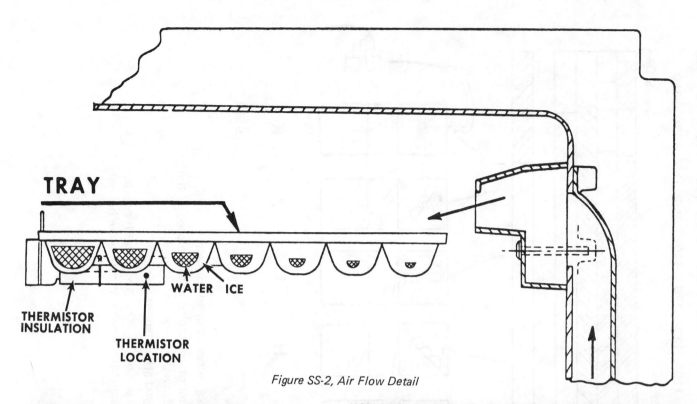

TRAY

THERMISTOR
INSULATION

THERMISTOR
LOCATION

WATER   ICE

Figure SS-2, Air Flow Detail

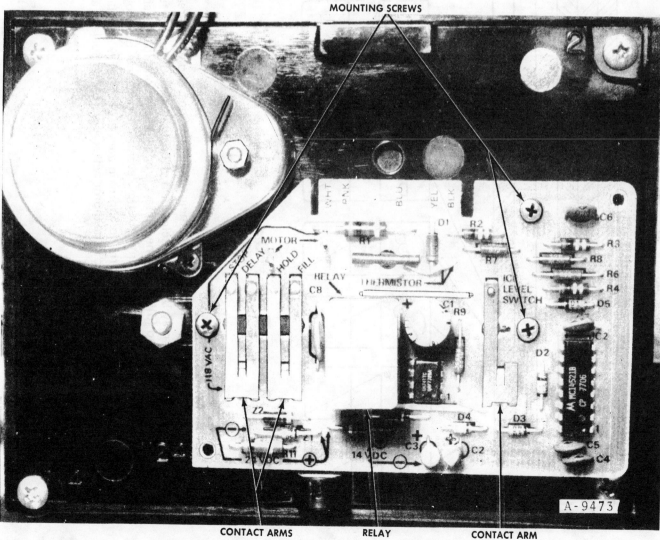

MOUNTING SCREWS

CONTACT ARMS   RELAY   CONTACT ARM

A-9473

Figure SS-3, Printed Circuit Board

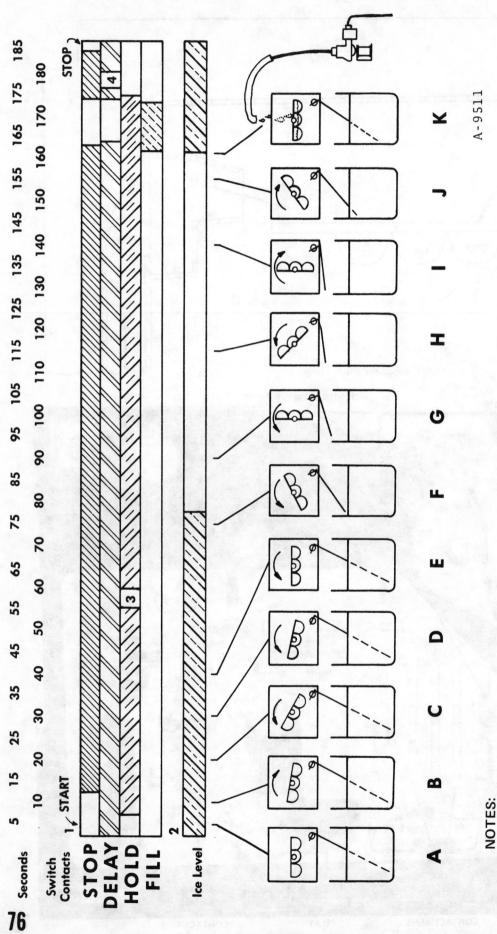

NOTES:

1. Shaded areas represent contacts in their closed positions, and white areas represent open contact.

2. Ice level contact may be open or closed. If the container is in and not full of ice, the contact is closed. If the container is in and full of ice, the contact is open. If the container is removed the contact is open.

3. Ice maker will stop at this point with tray level if the container is removed after Harvest starts.

4. If the thermistor did not warm to approximately +34° F. after the fill period: A—The No. 1 (cold) contact in the relay will remain closed. The motor will stop when the "delay" contact opens terminating the harvest cycle.

A-9511

Figure SS-5, The ICE-1 — Control Sequence

CONTROL CIRCUIT OPERATION
NORMAL AND "BACK UP" (TIMED) CYCLES

In normal operation the frequency of harvest cycles is dependent only upon the freezing capacity of refrigerator. However, a "back-up" timed freeze period is built into the control system to prevent harvesting of partially frozen cubes or malfunction of the ice maker under some abnormal conditions. This timed cycle is controlled by a digital timer which is incorporated into the solid state circuitry. If cubes in the front cavities fail to eject, ice forming in the front of the tray may be fully frozen and harvesting temperature achieved at the control thermistor while there is still some free water in the rear cavities of the tray. Under these conditions, the controls of the ICE-1 automatically initiate a timed freeze period which prevents harvesting of partially frozen cubes.

The timed freeze period will also be initiated in the event that water does not enter the tray during the fill period. Therefore, the ICE-1 will resume normal operation after the timed cycle expires, following temporary loss of water supply.

THIS IS HOW THE CONTROLS OPERATE.

NORMAL HARVEST CYCLE CONTROL

Refer to Electrical Schematic, Figure SS-4, and Control Sequence Chart, Figure SS-5.

During the normal freeze period the No. 2 (warm) contact of the single pole double throw relay is closed and the No. 1 (cold) contact open. When the thermistor senses approximately +13° F., indicating that the cubes are fully frozen, the No. 1 contact closes in series with the "Delay" contact. This completes the circuit to the motor and thus initiates the harvest cycle.

Contact No. 1 remains closed until the thermistor senses a temperature rise to approximately + 34° F. This occurs normally when water enters the tray during the "fill" period near the end of the harvest cycle. Within approximately 2 minutes after fill, contact No. 2 closes. This completes two circuits.

The circuit to the motor is completed through relay contact No. 2 until the "Stop" contact opens, terminating the harvest cycle, at which time a circuit is completed through relay contact No. 2 which stops the digital timers, this allows the subsequent harvest cycle to be temperature controlled through the thermistor. When the temperature at the thermistor is again reduced to approximately + 13° F., the relay opens contact No. 2 and closes No. 1 in series with the "Delay" contact, thus initiating the subsequent harvest cycle.

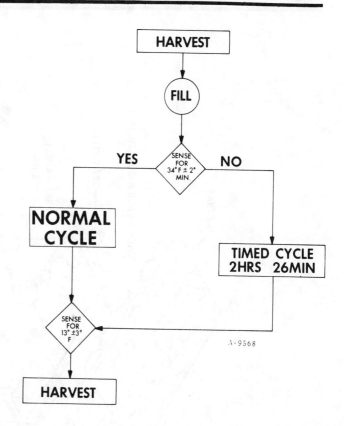

Figure SS-6, The ICE-1 — Operational Diagram

"BACK UP" (TIMED) HARVEST CYCLE CONTROL, Figure SS-6

If the temperature at the location of the thermistor is not warmed to approximately + 34° during the fill period, the No. 1 (cold) contact in the relay remains closed, leaving contact No. 2 (warm) open. With relay contact No. 1 closed after the fill period the harvest cycle is terminated by opening the "Delay" contact. This occurs a few seconds before the "Stop" contact opens.

Since relay contact No. 2 remains open under this condition the circuit to the digital timer in the electronic control circuit is open. This allows the timer to run thus initiating a "Timed Freeze Cycle" of 2 hours 26 minutes duration.

At the end of 2 hours, 26 minutes the digital timer in conjunction with the electronic control circuit transfers the relay contacts, opening contact No. 1 and closing contact No. 2. This completes a circuit through contact No. 2 and the "Stop" contact to the motor. The motor runs approximately 6 seconds until the "Stop" contact opens at the normal termination of the harvest cycle. At this point, provided that the thermistor senses approximately + 13° F., relay contact No. 1 recloses in series with the "Delay" contact, which initiates the harvest cycle.

In the event that the temperature at the location of the thermistor is above approximately + 13° F at the end of the 2 hours, 26 minutes Timed Freeze Cycle

77

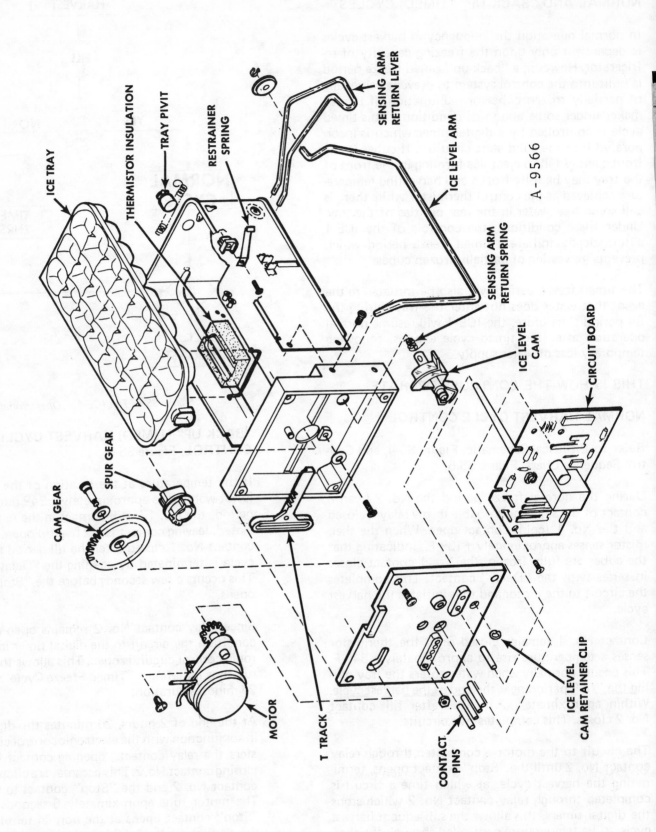

A - 9566

Figure SS-7, The ICE-1 Ice Maker — Exploded View

relay contact No. 2 remains closed completing the circuit that stops the digital timer. Thus when the thermistor senses a reduction in temperature to approximately + 13° F, relay contact No. 1 closes in series with the "Delay" contact, which immediately initiates the harvest cycle.

The control circuit sequence of operation for the timed harvest cycle is the same as for the normal harvest cycle after the cycle is initiated. The operation of the Ice Maker will return to normal for subsequent harvest cycles provided that the abnormal conditions are cleared at the termination of the timed cycle.

MECHANICAL OPERATION,
Refer to Figure SS-7 for Parts Identification

In operation, the motor drives the cam gear. An "off center lug" on the cam gear, Figure SS-8, fits into the slot at the end of the "T" rack. As the cam gear rotates this lug drives the "T" rack horizontally.

At the start of a harvest cycle, the "T" rack is moved to the left. A gear tooth on the end of the "T" rack engages the spur gear, rotating it clockwise. The spur gear is connected to the ice tray so the front of the tray, which is in the horizontal position at the start of a harvest cycle, Figure SS-5 (A), is twisted clockwise while the rear of the tray is held in place, Figure SS-5 (B). This twisting action frees the ice cubes within the cavities of the tray.

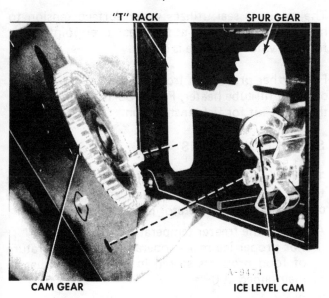

"T" RACK   **SPUR GEAR**

CAM GEAR   **ICE LEVEL CAM**

Figure SS-8, Cam Gear, "T" Rack Assembly Detail

As the cam gear continues to rotate, the "off center lug" moves the "T" rack to the right and the tray returns to the horizontal position, Figure SS-5 (C), (D) and (E).

The "T" rack continues to move to the right and a set of gear teeth near the center of the rack, Figure SS-8, engage the spur gear rotating the gear and the ice tray counterclockwise.

At this point, the end of the "T" rack comes into contact with a lever molded onto the ice level cam, Figure SS-9, rotating the cam clockwise. The ice level arm is mounted to the rear of the cam, Figure SS-10; so, as the cam rotates, the arm is lifted out of the ice container.

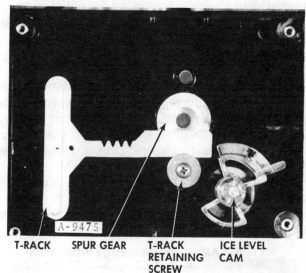

T-RACK   **SPUR GEAR**   **T-RACK RETAINING SCREW**   **ICE LEVEL CAM**

Figure SS-9, "T" Rack, Ice Level Cam Relationship

The "T" rack continues to move to the right until the ice tray is rotated several degrees past the vertical position, Figure SS-5 (F), (G) and (H), dropping cubes from the tray into the ice container. At this point the rear edge of the tray comes into contact with a restrainer spring which is mounted on the inside of the ice maker frame, Figure SS-11. This spring resists further rotation of the tray; then, due to its formation, releases the tray with a snap. This snap action jars the tray, rejecting any cubes that may not have fallen from the tray by their own weight. The "T" rack then moves to the left again, rotating the spur gear and the tray clockwise; Figure SS-5 (I), (J). As the gear teeth on the "T" rack disengage the spur gear the ice tray stops in the horizontal position, Figure SS-5 (K).

Simultaneously, the ice level cam, Figure SS-9, follows the movement of the "T" rack, rotating counterclockwise. Thus the ice level arm is lowered until it comes to rest on top of the ice in the container. The cam gear continues to rotate, energizing the water valve solenoid to fill the ice tray for the next cycle. The motor stops in the "Home" Tray Level position for the subsequent "Freeze" period.

The ice level arm return lever is not designed to raise the ice level arm independently. The ice level arm is moved upward by the rear of the ice container as the ice container is being removed. The return lever simply holds the arm in the "up" position to hold the ice level switch open.

As the ice container is replaced, the rear of the container comes into contact with the return lever moving it to the rear and releasing the ice level arm.

ICE MAKER AIR FLOW

Cold flowing air is delivered to the ice maker through a discharge duct in the air deflector at the rear of the freezer, Figure SS-2.

This duct is designed specifically to direct an even flow of cold air across the top of the entire width of the tray from the rear toward the front. This is essential to the production of good quality ice and reliable cube harvesting.

The thermistor which is utilized in the harvest control circuit senses the temperature of the center cube cavity in the second row from the front of the tray. Due to the direction of air flow, cubes in the front of the tray are the last to freeze. Thus it is assured that all cubes are fully frozen before the harvest cycle is initiated. The bottom of the tray is insulated at the location of the thermistor, Figure SS-10, so the harvesting controls are not influenced by the temperature below the tray.

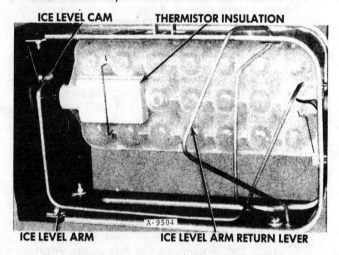

ICE LEVEL CAM   THERMISTOR INSULATION

ICE LEVEL ARM   ICE LEVEL ARM RETURN LEVER

Figure SS-10, Ice Level Arm Detail

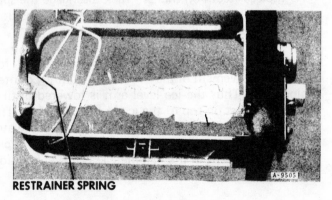

RESTRAINER SPRING

Figure SS-11, Restrainer Spring Detail

Almost all of the freezing air is directed over the top of the tray. Thus cubes are frozen from the top downward. As freezing progresses, thicker ice forms at the top of the cube than around the periphery, Figure SS-12. As the last small amount of water near the bottom of the cube freezes the force of the resultant expansion of the water follows the path of least resistance, cracking the periphery of the cube along the bottom. This forces the cube upward, releasing it from the cavity.

WATER SUPPLY SYSTEM

Water is supplied to the ice maker by a solenoid operated water valve which is mounted in the refrigerator machine compartment. Water flow through the valve is controlled by a flow washer. The flow washer is designed to pass 95 to 140 cc of water through the valve in 13 seconds, the normal fill period, when the flowing pressure of the water supply is between 20 and 120 P.S.I. at the inlet of the valve.

A plastic fill tube is connected between the outlet of the water valve and the inlet of a fill tube, Figure SS-13 which is factory installed in all models designed for ice maker installation.

A miniscus seal, Figure SS-13, is factory installed in the fill tube to prevent water from dripping and to assist in preventing it from freezing in nozzle outlet of the tube after the fill period.

The miniscus seal utilizes water surface tension to hold the water column at the break over point of the fill tube when the valve is closed.

For further protection from freezing at the fill tube outlet, a fill tube heater, Figure SS-18, is provided in the ICE-1 kit for field installation.

SERVICE DIAGNOSIS

The first step in diagnosing any problem in relation to the proper operation of the Ice Maker is to make certain that the refrigerator is operating properly.

A. Check the freezer temperature.
   For proper ice maker operation the temperature of food products stored in the freezer must be maintained at $0^o = 5^o$ F.

B. Check for proper air flow over the ice maker tray.
   1. Is the air deflector positioned properly and retained by screws? Figure SS-19.
   2. Is the freezer liner-to-air duct retaining screw, Figure SS-19, in place and tight?

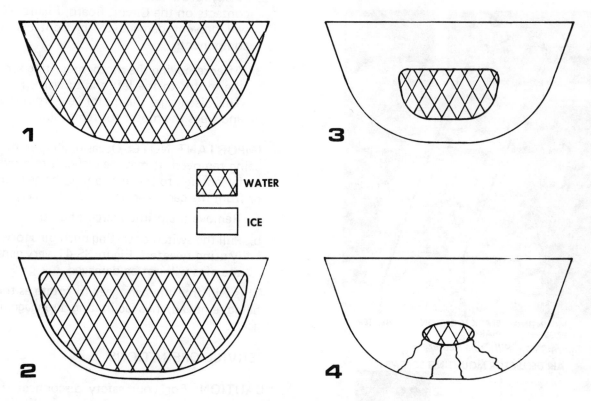

Figure SS-12, Progression Of Cube Freezing

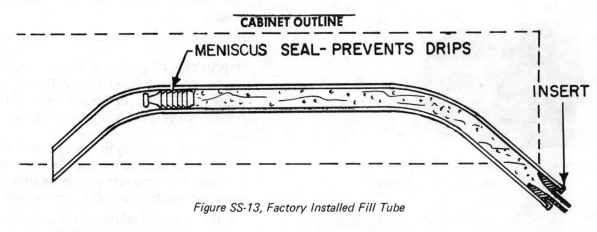

Figure SS-13, Factory Installed Fill Tube

3. Is the freezer fan motor running up-to-speed?

4. Is the evaporator relatively frost free so there is a minimum restriction to air flow through the evaporator?

SERVICE DIAGNOSIS CHECK-OUT PROCEDURES

The Diagnosis Check-Out Procedures are in the following order:

a. Ice Condition — Figure SS-20

b. No Ice Production — Figure SS-21

c. Low Ice Production — Figure SS-22

d. Electrical — Figure SS-23

e. Water Fill System — Figure SS-24

These procedures are designed to pinpoint problems through a series of tests. Proceed as follows:

1. Follow the "YES" arrow when you obtain the correct measurement or condition.

2. Follow the "NO" arrow when the correct measurement or condition is not obtained.

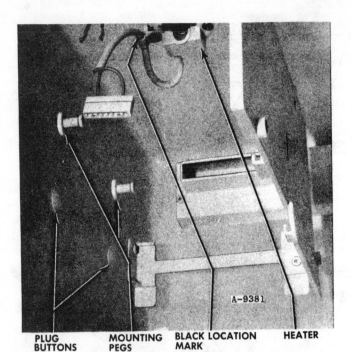

A-9381

PLUG   MOUNTING   BLACK LOCATION   HEATER
BUTTONS   PEGS   MARK

Figure SS-18, Fill Tube Heater

AIR DEFLECTOR MOUNTING SCREWS (2)

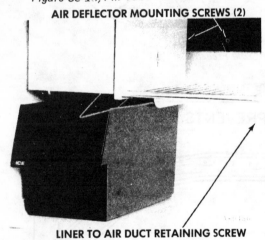

LINER TO AIR DUCT RETAINING SCREW

Figure SS-19, Freezer Detail — T Models

TROUBLESHOOTING PRECAUTIONS AND NOTES

WARNING: The full AC line voltage is present at several places on the Control Circuit Board. Be careful to avoid electrical shock when working with the Circuit Board.

1. Be careful to avoid shorting any terminals to ground when making voltage measurements.

2. Always use a non metallic probe to close the open contacts on the Circuit Board, Figure SS-25. This will prevent damage to the contacts and other components.

3. After removing the motor mounting screws, move the motor to the left and upward, Figure SS-40. This disengages the motor gear from the cam gear and frees the motor for removal.

IMPORTANT NOTE: Occasionally, as the motor is being removed, it may be difficult to move the motor far enough to the left to free the gear on the motor from the cam gear. In these instances:

a. Remove the printed circuit board.

b. Pull the switch operating push pins forward slightly using tweezers, Figure SS-41, very gently so that the pins will not be damaged.

This will free the cam gear to rotate as the motor is being removed, thus facilitating disengagement of the gears.

SERVICE OPERATIONS

CAUTION: For your safety disconnect the Refrigerator Electrical Supply Cord at the Wall Outlet before performing any service operation on the Ice Maker.

See Figure SS-53 for assembly sequence of all parts.

IMPORTANT: Ice Maker Motor—Replacement, Procedure C and Printed Circuit Board—Replacement, Procedure E may be conveniently performed with the ice maker installed in the refrigerator.

Due to the simplicity of removal and reinstallation it is more practical to remove the complete ice maker assembly from the refrigerator for most other service provided that one important precaution is observed:

Do not apply pressure to the printed circuit board or any of its components during removal or installation of the ice maker.

A. ICE MAKER, COMPLETE MECHANISM — removal and replacement

A. ICE MAKER, COMPLETE MECHANISM — REMOVAL AND REPLACEMENT

To remove the Ice Maker from the refrigerator:

1. Remove the ice maker front cover by removing the mounting screw, Figure SS-36, and pulling the cover outward at the bottom.

2. Disconnect the edge connector from the printed circuit board, Figure SS-37.

3. Remove the green "ground" lead from the motor.

4. Remove the ice maker mounting screws, Figure SS-38.

5. Pull straight outward on the ice maker assembly, to free it from the mounting pegs at the left hand side of the ice maker frame, Figure SS-36.

IMPORTANT REASSEMBLY NOTE:

a. To reinstall the terminal in the insulator, hold the insulator with the slotted holes at the top of the side facing you and insert the terminal with the locking tab at the top of the insulator, Figure SS-42.

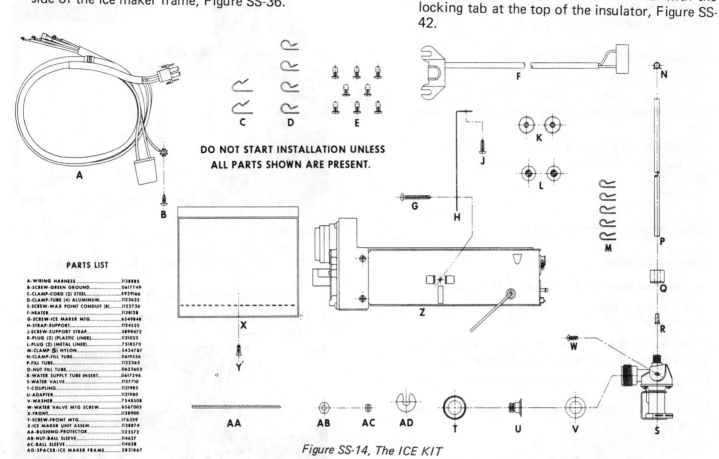

PARTS LIST

A-WIRING HARNESS..................1138885
B-SCREW-GREEN GROUND.............0617749
C-CLAMP-CORD (2) STEEL............5929166
D-CLAMP-TUBE (4) ALUMINUM.........1123635
E-SCREW-WAX POINT CONDUIT (8).....1127736
F-HEATER..........................1138128
G-SCREW-ICE MAKER MTG.............6549848
H-STRAP-SUPPORT...................1124525
J-SCREW-SUPPORT STRAP.............5899472
K-PLUG (2) PLASTIC LINER..........1121023
L-PLUG (2) METAL LINER............7518570
M-CLAMP (5) NYLON.................5434787
N-CLAMP-FILL TUBE.................0619556
P-FILL TUBE.......................1122363
Q-NUT FILL TUBE...................0623603
R-WATER SUPPLY TUBE INSERT........0617296
S-WATER VALVE.....................1137710
T-COUPLING........................1121985
U-ADAPTER.........................1121980
V-WASHER..........................7548508
W-WATER VALVE MTG SCREW...........6567005
X-FRONT...........................1138900
Y-SCREW-FRONT MTG.................176359
Z-ICE MAKER UNIT ASSEM............1138874
AA-BUSHING-PROTECTOR..............1123572
AB-NUT-BALL SLEEVE................114627
AC-BALL SLEEVE....................114628
AD-SPACER-ICE MAKER FRAME.........2831867

DO NOT START INSTALLATION UNLESS ALL PARTS SHOWN ARE PRESENT.

Figure SS-14, The ICE KIT

B. CONTROL, FRONT COVER — REMOVAL AND REPLACEMENT

To gain access to the control mounting plate, remove the ice maker front cover, Figure SS-36. To replace the cover position the wiring harness and fill nozzle heater leads as shown in Figure SS-39. Insert the tab at the top of the cover into the notch on the control mounting plate.

C. MOTOR REPLACEMENT

1. Disconnect the motor leads from the printed circuit board by pulling straight outward on the terminal insulators.

2. Disconnect the green ground lead from the motor terminal.

3. After removing the motor mounting screws, move the motor to the left and upward, Figure SS-40.

D. MOTOR TERMINAL INSULATOR — REPLACEMENT

1. Release the terminal locking tab from the insulator using a thin flat tool such as a ladies hair pin as shown in Figure SS-42.

b. Pull gently on the motor lead to be certain that the terminal is locked into the insulator.

E. PRINTED CIRCUIT BOARD — REPLACEMENT

Handle the circuit board by the edges. Do not apply pressure to the circuit board or any of its components.

1. Disconnect the motor leads from the printed circuit board by pulling straight outward on the terminal insulators.

2. Disconnect the terminal edge connector by pushing straight upward as shown in Figure SS-37.

3. After removing the board mounting screws pull straight outward on the board approximately 4'' until the thermistor probe mounted on the rear of the board clears the control mounting plate.

NOTE: Be sure that all three printed circuit board mounting screws are firmly tightened, but not overtightened. In the event that the threads in any one of the screw holes in the control mounting plate are stripped, replace the control mount-plate.

Proper operation of the electrical contacts on the printed circuit board is dependent upon firm re-retention of the board to the mounting plate.

F. PUSH PIN (SWITCH OPERATING) — REPLACEMENT

1. Remove the printed circuit board, Procedure E,

2. Pull the push pins out of their mounting holes in the control mounting plate using tweezers, very gently, Figure SS-43. so the pins will not be damaged.

   IMPORTANT REASSEMBLY NOTE: Lubricate the push pins lightly, using Sil-Glide, Part No. 5878010.

G. CAM GEAR — REPLACEMENT

1. Remove the motor, Procedure C.

2. Remove the printed circuit board, Procedure E.

3. Remove the retainer from the ice level cam shaft, Figure SS-44.

4. Remove the control plate mounting screws and separate the control plate from the gear housing.

5. Remove the cam gear mounting nut, Figure SS-44.

   NOTE: The cam gear mounting nut has left-hand threads.

   NOTE:

   a. Lightly lubricate the cams on the face of the cam gear, the off center lug on the opposite side of the cam gear, Figure SS-45 and the recess at the mounting screw location, Figure SS-46 using Sil-Glide Part No. 5787010.

   b. Install the cam gear on the control mounting plate, being certain that bearing washer, Figure SS-47, is assembled on the mounting screw.

   c. Use one drop of LocTite Part No. 6598018 on the threads of the cam gear mounting nut. Tighten the cam gear mounting nut firmly but do not overtighten.

   d. Check the cam gear to make certain that it rotates freely and that the end play between the gear and bearing washer is not excessive.

   e. Lightly lubricate the ice level cam and the bearing surface of the cam shaft which fits into the control plate, Figure SS-48, using Sil-Glide, Part No. 5878010.

   f. Reassemble the control mounting plate to the gear housing. BE SURE TO ALIGN THE OFF CENTER LUG ON THE CAM GEAR WITH THE SLOT IN THE END OF THE "T" RACK AND THE ICE LEVEL CAM SHAFT WITH ITS MOUNTING HOLE IN THE CONTROL MOUNTING PLATE AS SHOWN IN FIGURE SS-48.

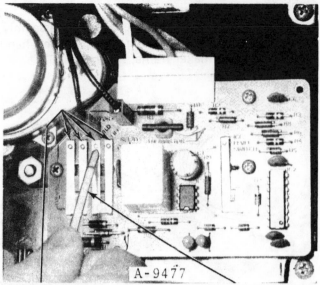

CONTACT ARMS   NON METALLIC PROBE

Figure SS-25, Closing "Hold" Contact

H. "T" RACK SPUR GEAR — REPLACEMENT

1. Remove the printed circuit board, Procedure E.

2. Remove the retainer from the ice level cam shaft, Figure SS-44.

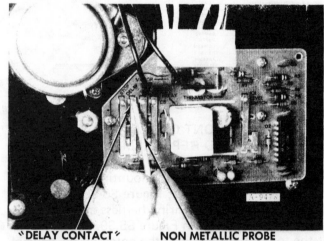

"DELAY CONTACT"   NON METALLIC PROBE

Figure SS-26, Closing "Delay" Contact

3. Remove the Control Mounting Plate mounting screws, Figure SS-44.

4. Remove the "T" rack spur gear retaining screw and washer, Figure SS-49. This frees the "T" rack for removal.

5. Pull straight outward on the spur gear to separate the hub of the gear from the tray and the rear housing.

   NOTE:

   a. Lightly lubricate the following points upon reassembly, using Sil-Glide Part No. 5878010. The bearing surface on the spur gear, Figure SS-50.

   The entire inside surface of the slot in the end of the "T" rack, Figure SS-51.

   The entire inside surface of the "T" rack guide slot in the gear housing, Figure SS-52.

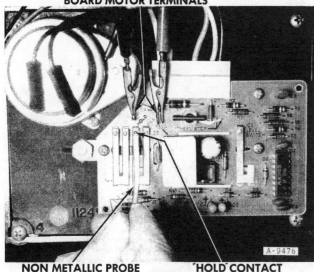

TEST LEADS CONNECTED TO CIRCUIT BOARD MOTOR TERMINALS

NON METALLIC PROBE   **"HOLD" CONTACT**

A-9476

Figure SS-27, Closing "Hold" Contact Without Test Leads

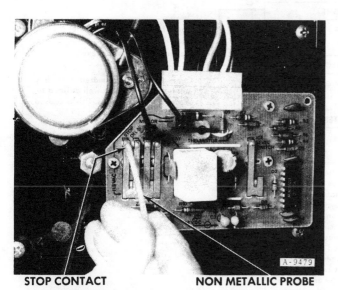

STOP CONTACT   **NON METALLIC PROBE**

A-9479

Figure SS-28, Closing "Stop" Contact

A-9480

PRINTED CIRCUIT BOARD MOUNTING SCREWS

Figure SS-29, 118 VAC Test — Circuit Board

B. Position the spur gear as shown in Figure SS-52, (ice tray level). Figure SS-52 also shows the proper assembly relationship between the gear tooth on the end of the "T" rack and the spur gear.

c. To reassemble the control mounting plate to the gear housing, BE SURE TO ALIGN THE OFF CENTER LUG ON THE CAM GEAR WITH THE SLOT IN THE END OF THE "T" RACK AND THE ICE LEVEL CAM WITH ITS MOUNTING HOLE IN THE CONTROL PLATE AS SHOWN IN FIGURE SS-48.

I. ICE LEVEL SENSING ARM — REPLACEMENT

1. Remove the sensing arm retainer, Figure SS-53.
2. Spring the sensing arm inward slightly to free it from the bearing in the ice maker frame, Figure SS-53A.
3. Rotate the sensing arm 90° to align the locking tabs with the slot in the ice level cam, Figure SS-54 to free the arm from the cam.

NOTE: Be sure to position the tang on the ice level arm return spring as shown in Figure SS-54.

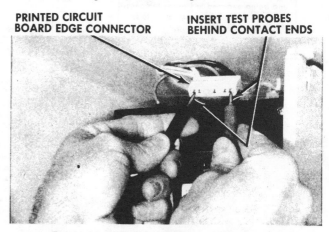

PRINTED CIRCUIT BOARD EDGE CONNECTOR   **INSERT TEST PROBES BEHIND CONTACT ENDS**

Figure SS-30, 118 VAC Test — Edge Connector

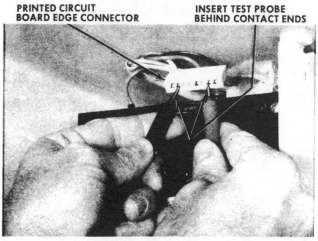

PRINTED CIRCUIT BOARD EDGE CONNECTOR   **INSERT TEST PROBE BEHIND CONTACT ENDS**

Figure SS-31, Resistance Check — Fill Tube Heater

ICE CONDITION CHECK LIST

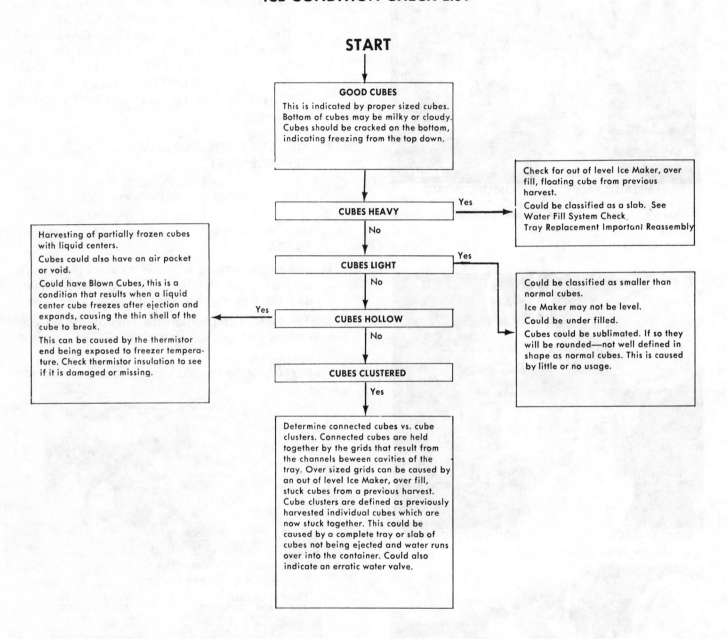

START

GOOD CUBES
This is indicated by proper sized cubes. Bottom of cubes may be milky or cloudy. Cubes should be cracked on the bottom, indicating freezing from the top down.

CUBES HEAVY — Yes →
Check for out of level Ice Maker, over fill, floating cube from previous harvest.
Could be classified as a slab. See Water Fill System Check.
Tray Replacement Important Reassembly

No

CUBES LIGHT — Yes →
Could be classified as smaller than normal cubes.
Ice Maker may not be level.
Could be under filled.
Cubes could be sublimated. If so they will be rounded—not well defined in shape as normal cubes. This is caused by little or no usage.

No

CUBES HOLLOW — Yes →
Harvesting of partially frozen cubes with liquid centers.
Cubes could also have an air pocket or void.
Could have Blown Cubes, this is a condition that results when a liquid center cube freezes after ejection and expands, causing the thin shell of the cube to break.
This can be caused by the thermistor end being exposed to freezer temperature. Check thermistor insulation to see if it is damaged or missing.

No

CUBES CLUSTERED

Yes

Determine connected cubes vs. cube clusters. Connected cubes are held together by the grids that result from the channels beween cavities of the tray. Over sized grids can be caused by an out of level Ice Maker, over fill, stuck cubes from a previous harvest. Cube clusters are defined as previously harvested individual cubes which are now stuck together. This could be caused by a complete tray or slab of cubes not being ejected and water runs over into the container. Could also indicate an erratic water valve.

Figure SS-20, Service Diagnosis — "Ice Condition" Check Procedures

NO ICE PRODUCTION

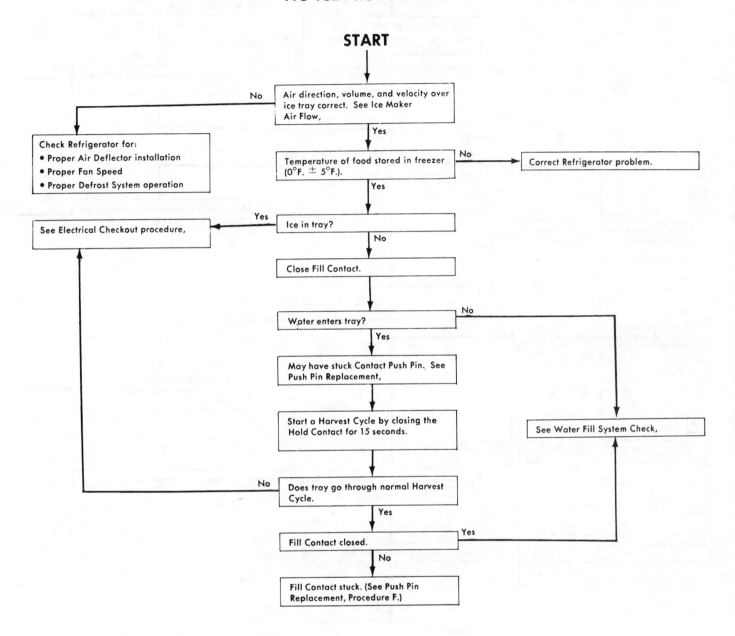

Figure SS-21, Service Diagnosis — "No Ice Production" Check Procedure

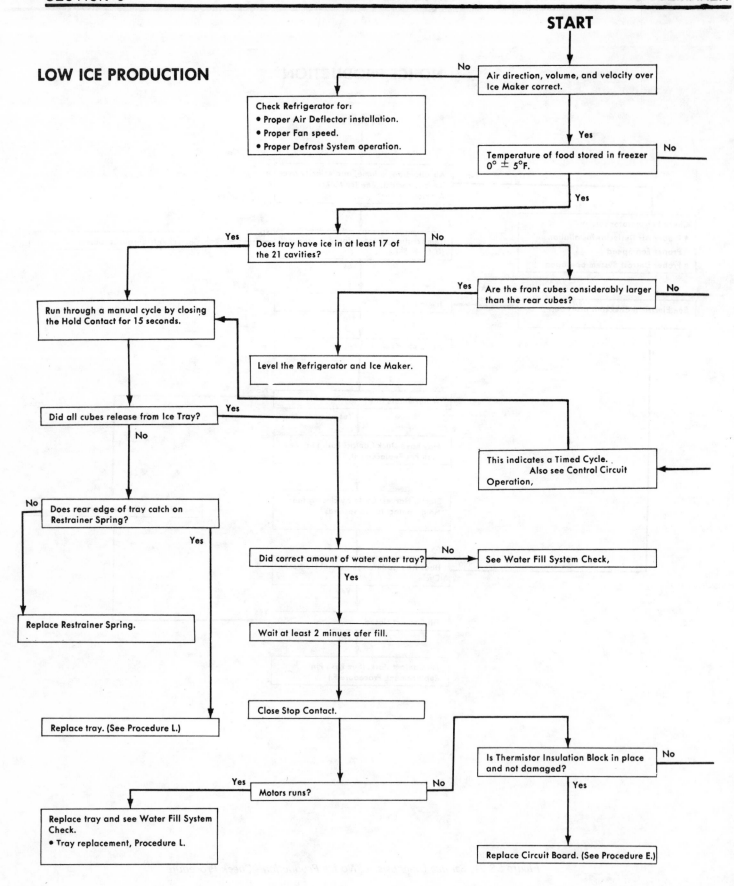

Figure SS-22, Service Diagnosis — "Low Ice Production" Check Procedure

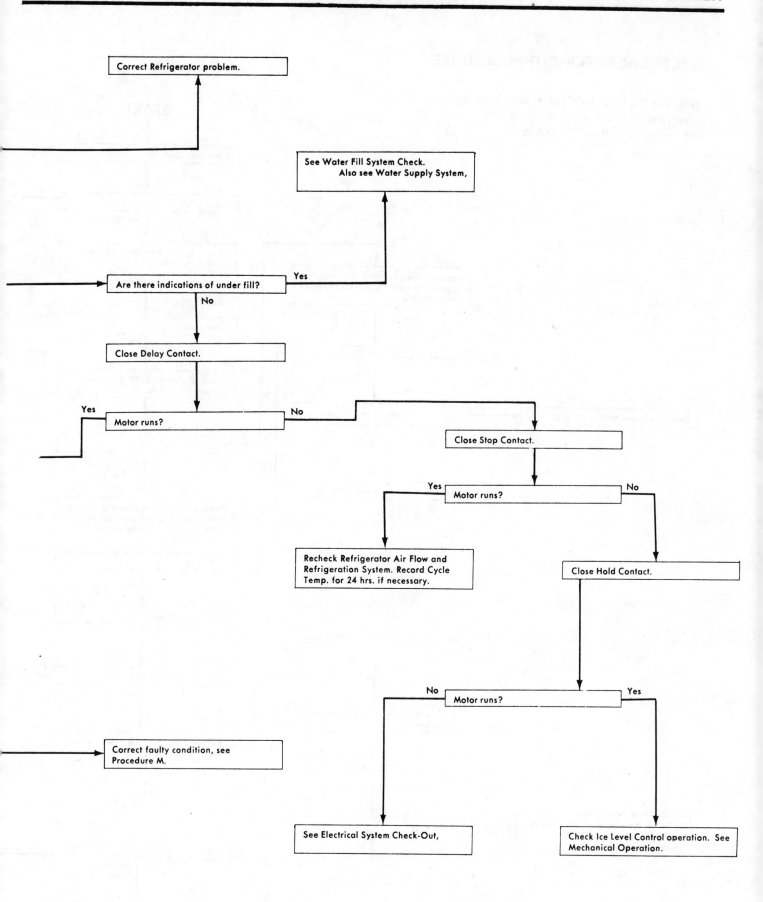

Figure SS-22, Service Diagnosis – "Low Ice Production" Check Procedure

ELECTRICAL CHECK-OUT PROCEDURE

Note: Use the Electrical Contact Circuits and Schematics to aid in understanding line voltage circuit on Circuit Board.

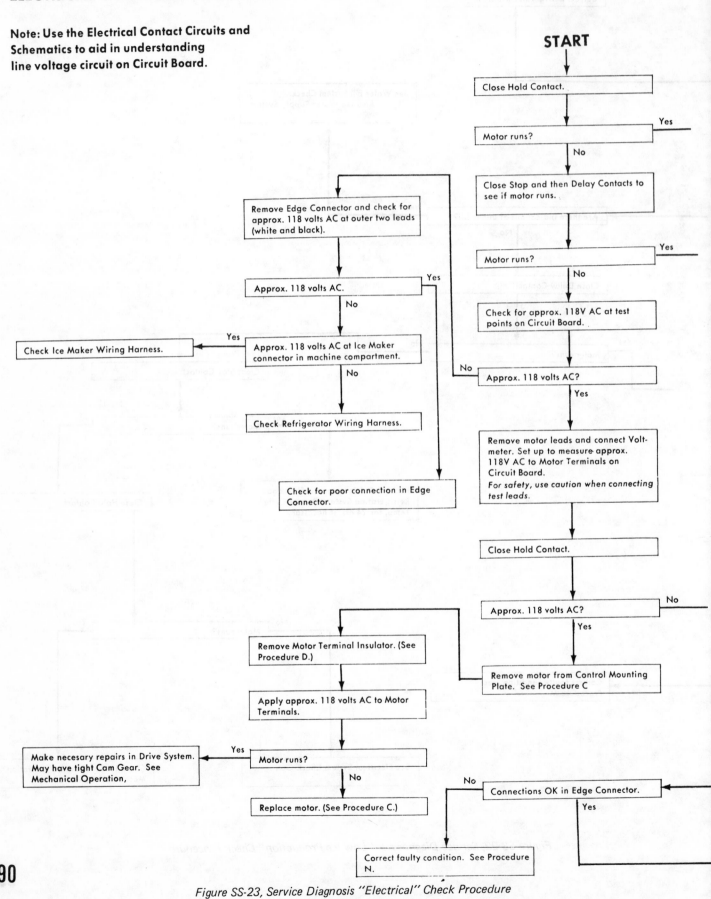

Figure SS-23, Service Diagnosis "Electrical" Check Procedure

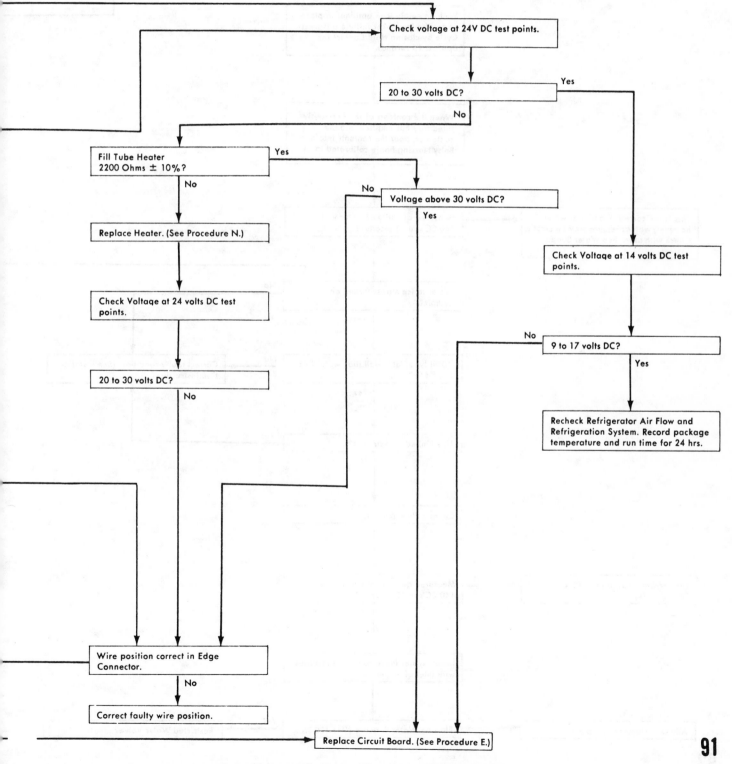

Figure SS-23, Service Diagnosis — "Electrical" Check Procedure

WATER SYSTEM CHECK-OUT PROCEDURE

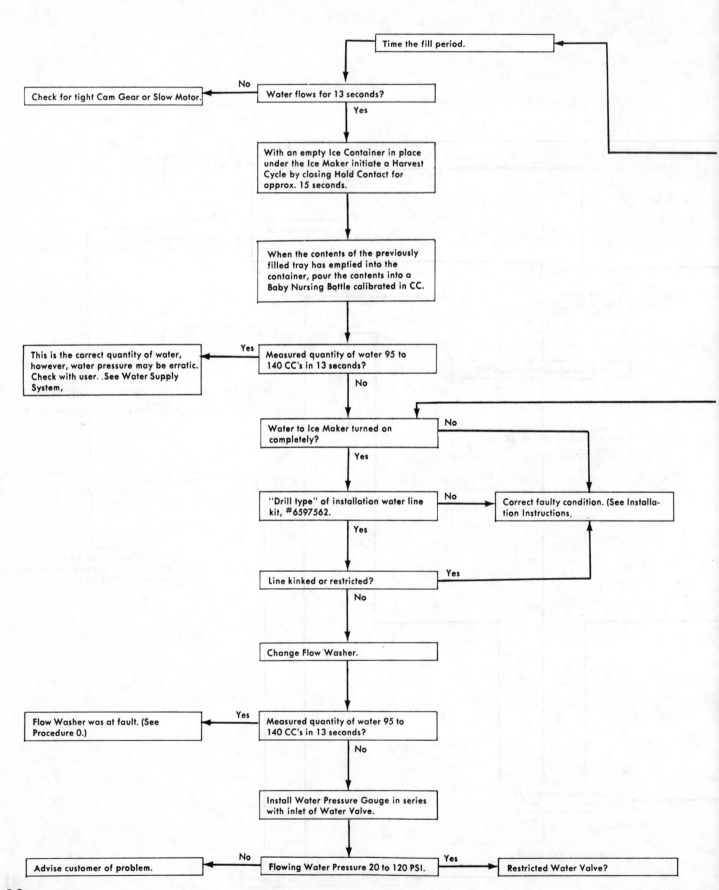

Time the fill period.

Water flows for 13 seconds? — **No** → Check for tight Cam Gear or Slow Motor.

Yes

With an empty Ice Container in place under the Ice Maker initiate a Harvest Cycle by closing Hold Contact for approx. 15 seconds.

When the contents of the previously filled tray has emptied into the container, pour the contents into a Baby Nursing Bottle calibrated in CC.

Measured quantity of water 95 to 140 CC's in 13 seconds? — **Yes** → This is the correct quantity of water, however, water pressure may be erratic. Check with user. See Water Supply System.

No

Water to Ice Maker turned on completely? — **No** → Correct faulty condition. (See Installation Instructions.

Yes

"Drill type" of installation water line kit, #6597562. — **No** → Correct faulty condition. (See Installation Instructions.

Yes

Line kinked or restricted? — **Yes** →

No

Change Flow Washer.

Measured quantity of water 95 to 140 CC's in 13 seconds? — **Yes** → Flow Washer was at fault. (See Procedure 0.)

No

Install Water Pressure Gauge in series with inlet of Water Valve.

Advise customer of problem. ← **No** — Flowing Water Pressure 20 to 120 PSI. — **Yes** → Restricted Water Valve?

Figure SS-24, Service Diagnosis — "Water Fill System" Check Procedure

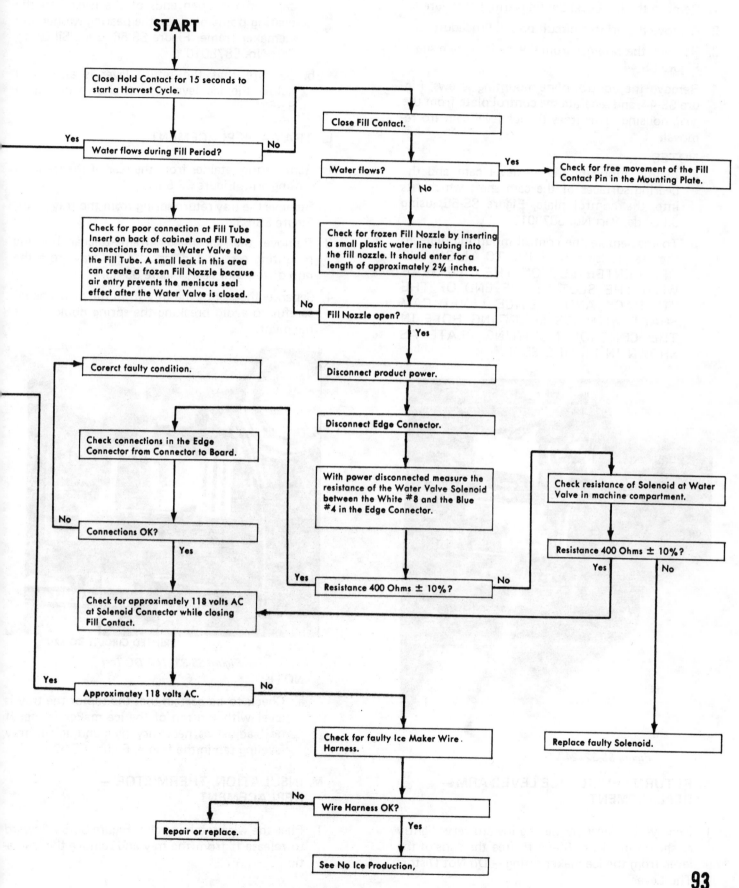

Figure SS-24, Service Diagnosis — "Water Fill System" Check Procedure

J. ICE LEVEL CAM — REPLACEMENT

1. Remove the Ice Level Sensing Arm, Procedure I.

2. Remove the printed circuit board, Procedure E.

3. Remove the retainer from the ice level cam shaft, Figure SS-44.

4. Remove the control plate mounting screws, Figure SS-44, and separate the control plate from the gear housing. This frees the ice level cam fro removal.

   NOTES:

   a. Lightly lubricate the ice level cam and the bearing surfaces of the cam shaft which fits into the control plate, Figure SS-55, using Sil-Glide, Part No. 5878010.

   b. To reassemble the control mounting plate to the gear housing, BE SURE TO ALIGN THE OFF CENTER LUG ON THE CAM GEAR WITH THE SLOT IN THE END OF THE "T" RACK AND THE ICE LEVEL CAM SHAFT WITH ITS MOUNTING HOLE IN THE CONTROL MOUNTING PLATE AS SHOWN IN FIGURE SS-48.

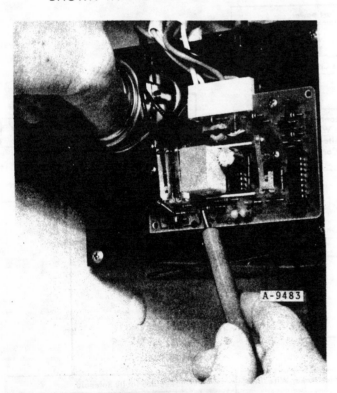

Figure SS-32, 24 VDC Test

K. RETURN LEVER — ICE LEVEL ARM — REPLACEMENT

1. Remove the lever by flexing inward very slightly as shown in Figure SS-54 to free the ends of the lever from the ice maker spring — Do Not Distort The Lever.

2. Thread the longer branch of the lever through the mounting hole to free it from the icemaker frame.

IMPORTANT REASSEMBLY NOTE:

a. Lightly lubricate the insides of the bearing caps on the open ends of the lever and the bearing point between the bearing washer and icemaker frame, Figure SS-56, using Sil Glide, Part No. 5878010.

b. Be sure that the return lever is assembled below the ice level arm as shown in Figure SS-53.

L. TRAY — REPLACEMENT

1. Remove the retainer from the rear of the ice level sensing arm, Figure SS-53.

2. Remove the tray return spring from the tray pivot, Figure SS-53.

3. Remove the ice maker frame-to-gear housing mounting screws and separate the tray from the hub of the spur gear, Figure SS-53.

4. Separate the rear of the tray from the pivot, being careful to avoid breaking the spring hook off of the pivot.

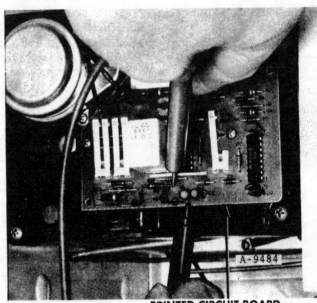

PRINTED CIRCUIT BOARD

Figure SS-33, 14 VDC Test

NOTE:

a. Check to be certain that the top of the tray is level with the top of the ice maker frame. If not, adjust as necessary by bending the tray leveling tab in the frame, Figure SS-57.

M. INSULATION, THERMISTOR — REPLACEMENT

1. Flex the wire retaining clip, Figure SS-54, inward to release it from the tray and remove the insulation.

IMPORTANT REASSEMBLY NOTE:

2. Be certain that the open end of the formation in the top of the insulation, is positioned in firm contact with the spur gear boss at the front of the tray.

3. Be certain that the wire insulation retainer is not bent or malformed and that it holds the insulation in firm contact with the bottom of the tray.

N. FILL TUBE HEATER AND EDGE CONNECTOR — REPLACEMENT

1. Remove the ice maker assembly from the refrigerator, Procedure A.

2. Remove the wiring harness leads, (black, blue and white) from the edge connector, Figure SS-58.

3. Remove and replace the heater as shown in Figure SS-59.

NOTE:

a. Route the heater wiring harness through the plastic retaining strap, observing the dimension shown in Figure SS-59 in forming the loop in the harness between the heater and the retaining strap.

b. Connect the wiring harness leads to the edge connector, following the connection diagram, Figure SS-60. Be sure to insert the terminals on the leads into the edge connector cavities with the locking tabs on the terminals located in the channel at the rear of the cavity as shown in Figure SS-60. It may be necessary to bend the locking tabs out slightly. Pull gently on the leads to be certain that the terminals lock into the edge connector.

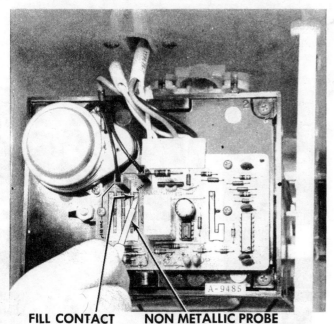

FILL CONTACT   NON METALLIC PROBE

Figure SS-34, Closing Fill Center

O. WATER VALVE SERVICE

Reference Figure SS-61 for assembly sequence of all water valve parts.

IMPORTANT REASSEMBLY NOTES

1. Flow washer replacement—The flow washer must be assembled in the valve inlet, Figure SS-62, with the lettered side of the washer facing outward.

2. Ice Maker Fill Tube Connection—Be certain that the red plastic insert is fully seated in the fill tube, Figure SS-63. The plastic nut should be finger tightened on the valve outlet.

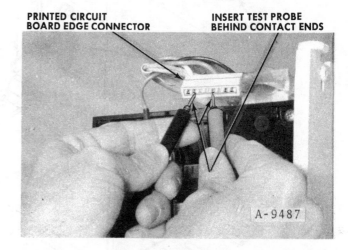

PRINTED CIRCUIT   INSERT TEST PROBE
BOARD EDGE CONNECTOR   BEHIND CONTACT ENDS

Figure SS-35, Resistance Check — Water Valve Solenoid

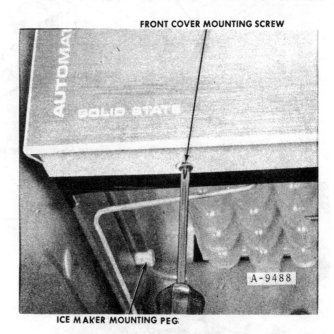

FRONT COVER MOUNTING SCREW

ICE MAKER MOUNTING PEG

Figure SS-36, Front Cover Removal

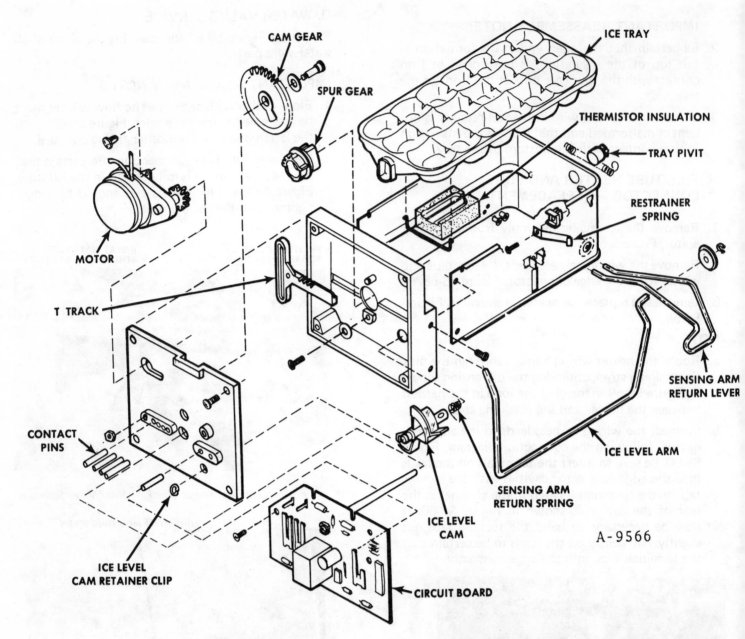

Figure SS-35A, The ICE-1 Ice Maker — Exploded View

Labels: CAM GEAR, SPUR GEAR, ICE TRAY, THERMISTOR INSULATION, TRAY PIVIT, RESTRAINER SPRING, MOTOR, T TRACK, SENSING ARM RETURN LEVER, CONTACT PINS, ICE LEVEL CAM, SENSING ARM RETURN SPRING, ICE LEVEL ARM, ICE LEVEL CAM RETAINER CLIP, CIRCUIT BOARD

A-9566

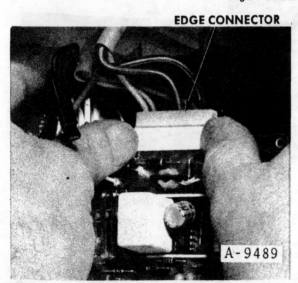

EDGE CONNECTOR

A-9489

96 *Figure SS-37, Removing Edge Connector*

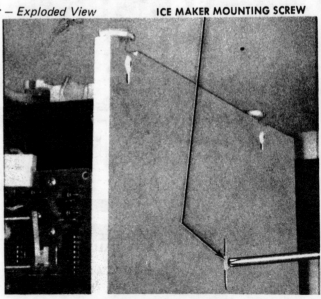

ICE MAKER MOUNTING SCREW

Figure SS-38, Ice Maker Mounting Screw — Removal

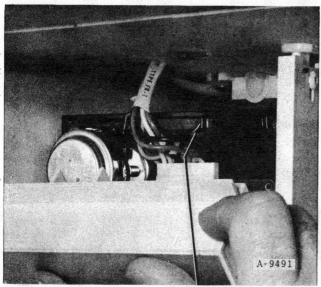

COVER MOUNTING NOTCH

Figure SS-39, Front Cover — Replacement

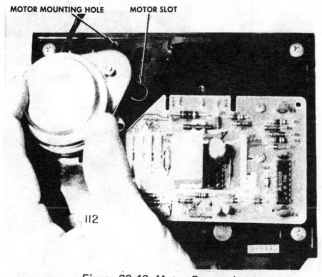

MOTOR MOUNTING HOLE   MOTOR SLOT

112

Figure SS-40, Motor Removal

SWITCH OPERATING PUSH PINS

A-9493

Figure SS-41, Disengaging Push Pins From Cam Gear

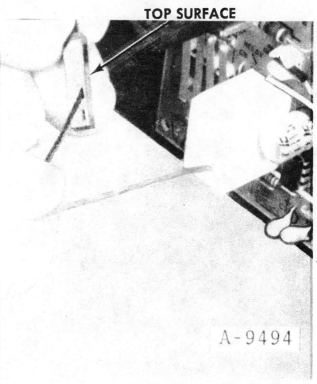

TOP SURFACE

A-9494

Figure SS-42, Motor Terminal Insulator — Removal

SWITCH OPERATING PUSH PINS

Figure SS-43, Switch Operating Push Pin — Removal

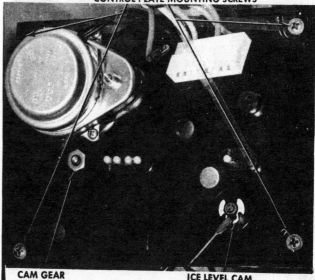

CONTROL PLATE MOUNTING SCREWS

CAM GEAR
MOUNTING NUT

ICE LEVEL CAM
SHAFT RETAINER

Figure SS-44, Ice Level Cam Shaft Retainer — Removal

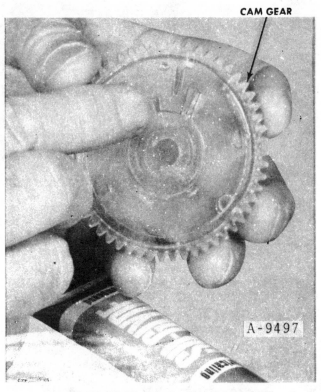

Figure SS-45, Lubrication of Cam Gear

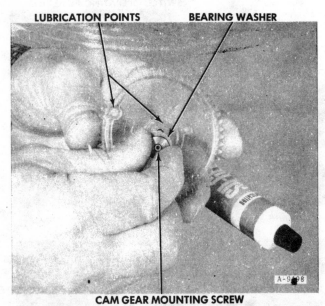

Figure SS-46, Cam Gear Bearing Washer Lubrication

Figure SS-47, Cam Gear Bearing Washer — Installed

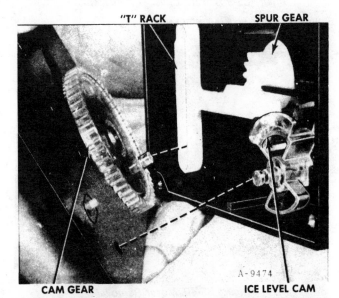

Figure SS-48, Cam Gear, "T" Rack Assembly — Detail

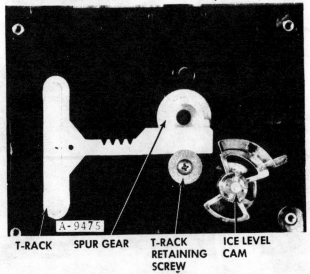

Figure SS-49, "T" Rack, Spur Gear Mounting — Detail

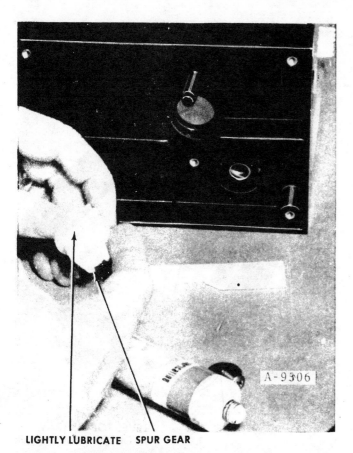

LIGHTLY LUBRICATE   SPUR GEAR

Figure SS-50, Spur Gear Lubrication

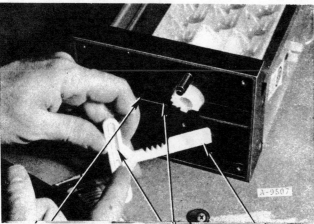

T-RACK SLOT   LIGHTLY LUBRICATE   T-RACK

Figure SS-51, "T" Rack Lubrication

"T" RACK GUIDE SLOT   SPUR GEAR

"T" RACK   ICE LEVEL CAM

Figure SS-52, "T" Rack, Spur Gear — Assembly Relationship

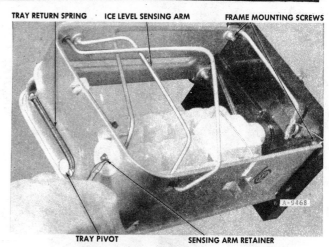

TRAY RETURN SPRING   ICE LEVEL SENSING ARM   FRAME MOUNTING SCREWS

TRAY PIVOT   SENSING ARM RETAINER

Figure SS-53, Ice Level Sensing Arm — Retainer

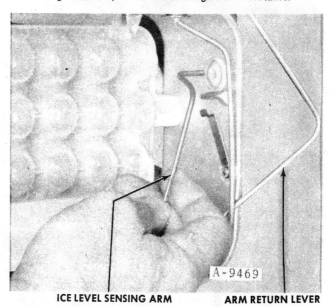

ICE LEVEL SENSING ARM   ARM RETURN LEVER

Figure SS-53A, Ice Level Sensing Arm — Removal

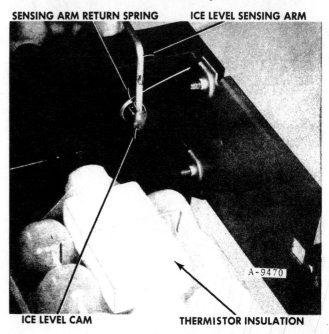

SENSING ARM RETURN SPRING   ICE LEVEL SENSING ARM

ICE LEVEL CAM   THERMISTOR INSULATION

Figure SS-54, Ice Level Arm-to-Cam Mounting — Detail

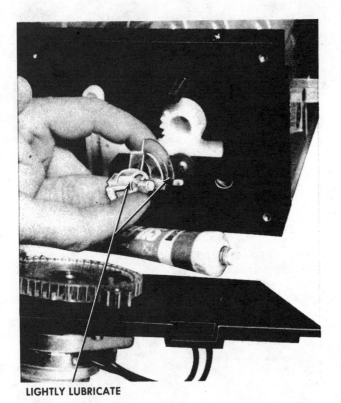

LIGHTLY LUBRICATE

Figure SS-55, Ice Level Cam Lubrication

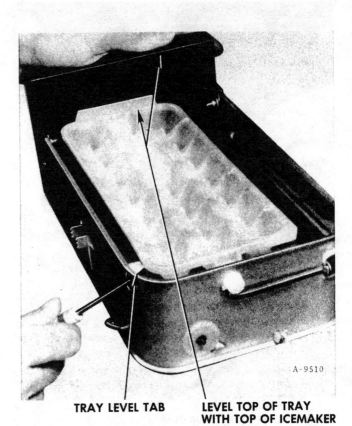

TRAY LEVEL TAB   **LEVEL TOP OF TRAY WITH TOP OF ICEMAKER**

Figure SS-57, Tray Level Adjustment

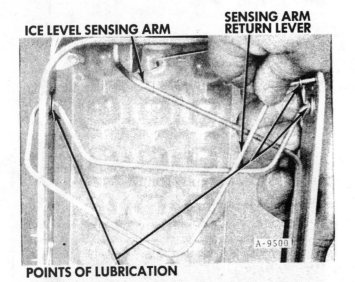

ICE LEVEL SENSING ARM   **SENSING ARM RETURN LEVER**

POINTS OF LUBRICATION

Figure SS-56, Ice Level Arm Return Lever — Removal

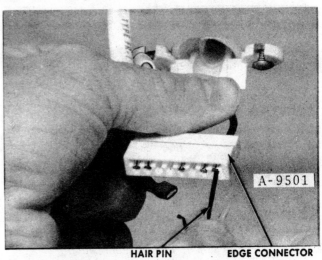

HAIR PIN   **EDGE CONNECTOR**

Figure SS-58, Edge Connector Terminal Removal

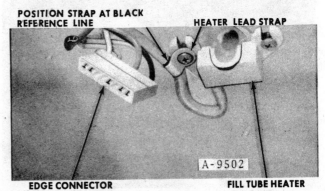

POSITION STRAP AT BLACK REFERENCE LINE   **HEATER LEAD STRAP**

EDGE CONNECTOR   **FILL TUBE HEATER**

Figure SS-59, Fill Tube Heater Installation Detail

WIRING HARNESS

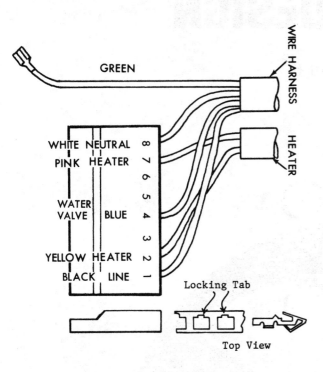

Figure SS-60, Wiring to Edge Connector — Connection Detail

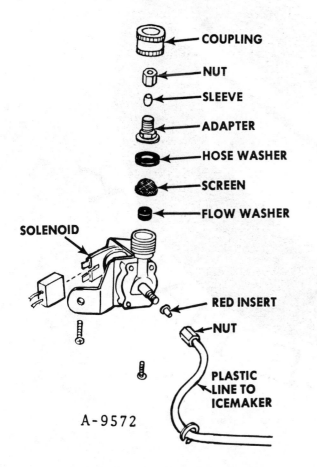

Figure SS-61, Water Valve — Exploded View

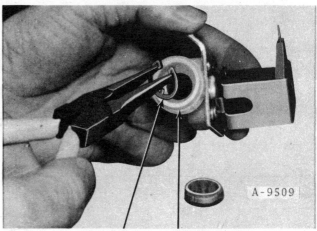

RETAINER CLIP   WATER VALVE
Figure SS-62, Flow Washer — Assembly Detail
Flow Washer Assembly Detail

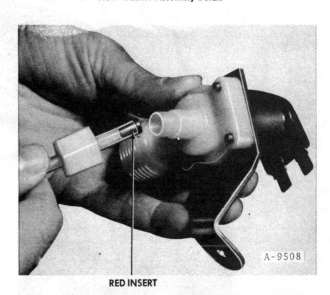

RED INSERT

Figure SS-63, Fill Tube to Water Valve — Connection Detail

FRIGIDAIRE DESIGN

IMK SERIES 1 thru 6

Be sure to satisfy essential requirements for ice production described on the previous page before using this chart.

HOW TO USE THIS CHART: Follow arrow to the right if statement in the block is true. Follow arrow downward if statement in the block is not true. Always choose proper option if given.

ICEMAKER WEIGH SYSTEM AND THERMOSTAT CHECKOUT PROCEDURE

START HERE

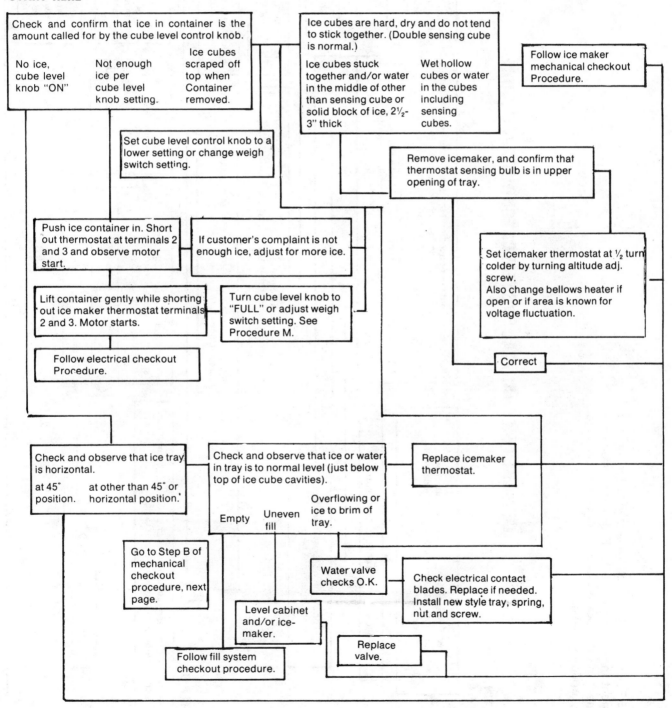

TO USE THIS CHART: Follow arrow to the right if statement in the block is true. Follow arrow down if statement in the block is not true. Always choose the proper option if given.

ICE MAKER MECHANICAL SYSTEM CHECKOUT PROCEDURE

CONTINUE HERE
STEP A

STEP B

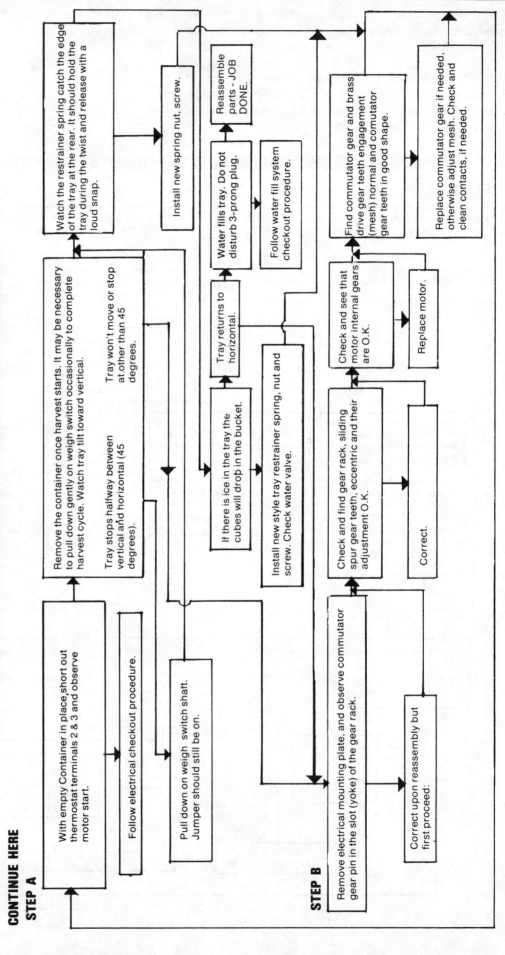

TO USE THIS CHART: Follow arrow to the right if statement in the block is true. Follow arrow down if statement in the block is not true. Always choose the proper option if given.

WATER FILL SYSTEM CHECK OUT PROCEDURE

CONTINUE HERE

ELECTRICAL CHECKOUT PROCEDURE

CONTINUE HERE

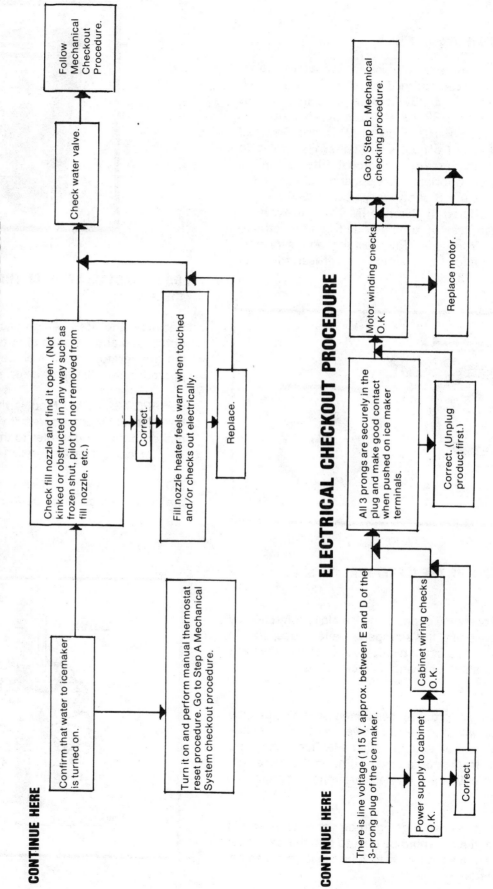

WATER FILL SYSTEM CHECK OUT PROCEDURE

Confirm that water to icemaker is turned on.

Turn it on and perform manual thermostat reset procedure. Go to Step A Mechanical System checkout procedure.

Check fill nozzle and find it open. (Not kinked or obstructed in any way such as frozen shut, pilot rod not removed from fill nozzle, etc.)

Correct.

Fill nozzle heater feels warm when touched and/or checks out electrically.

Replace.

Check water valve.

Follow Mechanical Checkout Procedure.

ELECTRICAL CHECKOUT PROCEDURE

There is line voltage (115 V. approx. between E and D of the 3-prong plug of the ice maker.

Power supply to cabinet O.K.

Correct.

Cabinet wiring checks O.K.

All 3 prongs are securely in the plug and make good contact when pushed on ice maker terminals.

Correct. (Unplug product first.)

Motor winding checks O.K.

Replace motor.

Go to Step B. Mechanical checking procedure.

FRIGIDAIRE IMK-1 ICE MAKER

INSTALLATION (Figure F1)

Proper installation of the ice maker water supply can be easily accomplished by choosing a kit such as the Gemline IM 251 (saddle valve installations on water lines 3/8″ to 1 3/8″). A charcoal line filter Gemline WF 101 (Figure F-2) is recommended between the shut off valve and water inlet valve to ice maker, the use of a charcoal filter assures a sediment free, odorless, tasteless and sparkling clear ice cubes. Figure 3 shows a typical water line connection as offered in Gemline IM 251. Always leave eight to ten feet of tubing coiled behind refrigerator away from wall for cleaning and servicing. Flush line before connection to refrigerator inlet valve. Figure F3.

Figure F-2

ADD ON INSTALLATION OF THE IMK FRIGIDAIRE ICE MAKERS

Because the IMK Automatic Ice Maker can be purchased and installed after the original purchase of the refrigerator, we will include the proper installation procedures of the ice maker, Figure F 4

For FPC 13-220 VU Models, there are exceptions to the installation procedures. If you are installing an ice maker in this model, refer to the instructions on page 99 for correct procedures.

NOTE: An asterisk after the procedure number, makes reference to procedures on page 93 thru 99

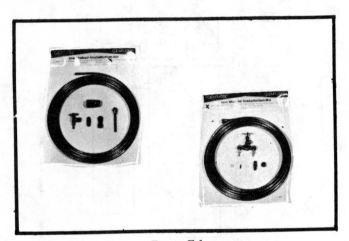

Figure F-1

PROPER USE OF THE ICE MAKER

The level of the ice cubes in the ice bin is controlled by actually weighing the storage bin and cubes, so it is very important that the bin be pushed all the way back into place after removing any ice cubes.

If the bin is not properly positioned on the guides, the ice maker will shut off at a lower level of ice cubes then the control setting. Whenever ice cubes are removed from the storage bin, make sure the remaining cubes are leveled throughout the bin.

Whenever the user is going to be gone for extended periods of time, it would be best to turn the ice maker off, and remove the ice bin to insure that ice cubes will not be made.

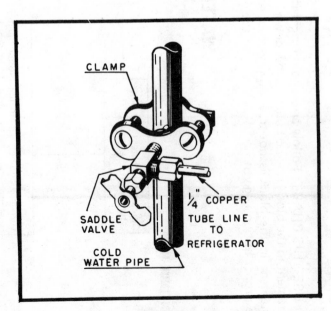

Figure F-3

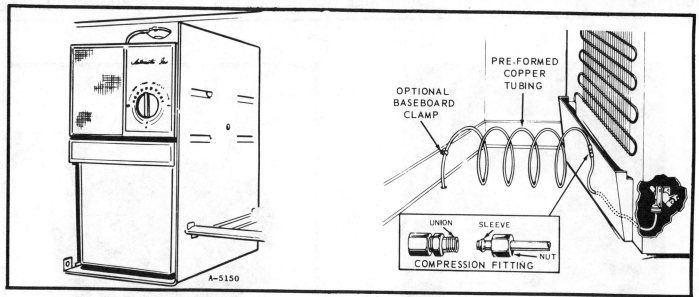

Figure F-4

INSTALLATION OF THE IMK-4 AUTOMATIC ICEMAKER

1. Lay out all parts in the kit to be certain that all parts
   are accounted for, Figure F5.

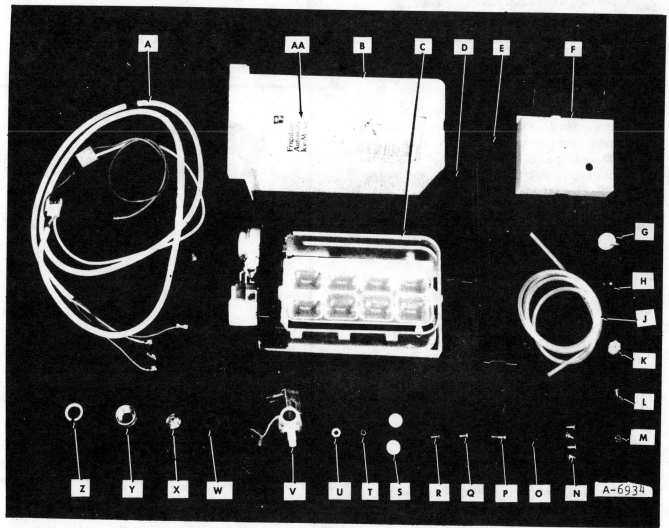

Figure F-5 PARTS BREAKDOWN IMK ICE MAKER

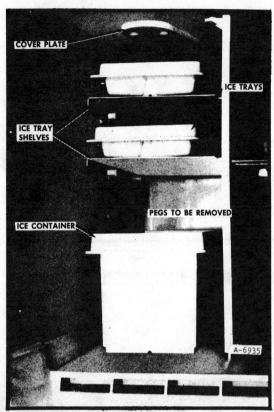

Figure F-6

ICE TRAYS. SHELVES, AND MOUNTING PEGS

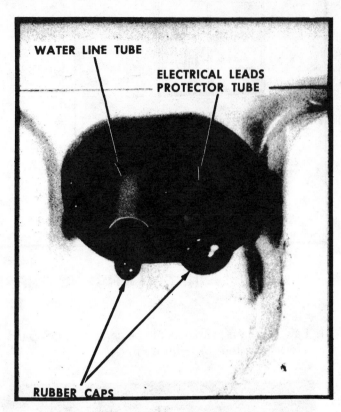

Figure F-7

PROTECTOR TUBES—REAR OF CABINET

2. Unplug the refrigerator.

3. Open the freezer door, remove and discard the ice trays, ice container, and ice tray shelves, Figure F6. With these parts removed, a label reading "Ice Maker Mounting Location" will be visible between the upper two pegs on the left side of the freezer, Figure F9.

4. Remove and discard the two lower pegs just below the two pegs marked "Ice Maker Mounting Location." A 90° to 180° clockwise twist with a wrench or pliers will loosen the pegs so that they can be pulled out, Figure F6.

5. Push the two white plug buttons (Part "S") into the two peg mounting holes, Figure F9.

6. Remove the four screws that mount the cover plate located in the top of the freezer liner, Figure F6. Discard the cover and reinstall the screws in the same holes.

7.* Move the refrigerator out to gain access to the rear. Remove the two rubber caps from the protector tube and the water line tube and discard them, Figure F7.

8. Start the electrical lead harness (Part "A") into the larger of the two tubes (one on the right as viewed from the rear). Align the leads and terminals, Figure F8, for easy feeding through the tube into the freezer compartment.

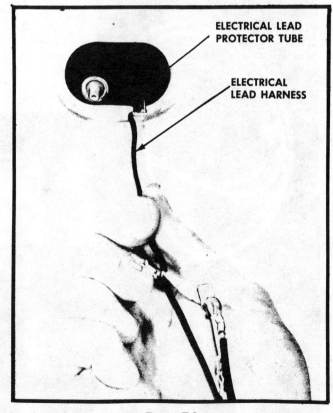

Figure F-8

ELECTRICAL LEAD HARNESS INSTALLATION

9. Install the electrical terminals into the 3-prong plug, (Part "H") following the color coding of the plug. When properly assembled, the tabs on the terminals lock into position inside the 3-prong plug to prevent accidental removal, Figure F9. If the terminals do not lock in place, bend the tab out slightly and reinsert into plug.

10. Observe the ice maker mechanism (Part "C"). The left side of the frame has a long slot, Figure F10. This slot slides onto the two pegs on the left side of the liner. The right side of the frame has three hooks. The front and rear hooks fit mating hooks on the vertical freezer divider. Slide the ice maker onto the pegs and engage the hooks. The electrical plug must be held up out of the way during the installation.

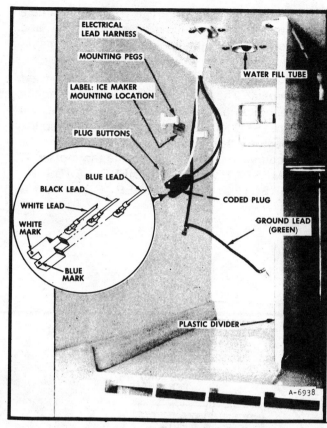

Figure F-9
ELECTRICAL LEAD HARNESS—FREEZER COMPARTMENT

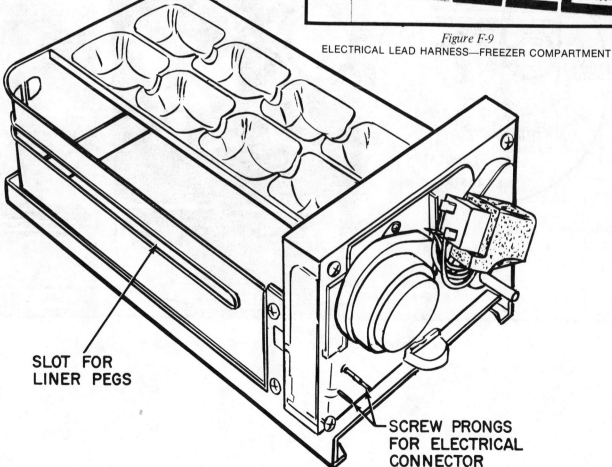

Figure F-10
ICE MAKER MECHANISM

11. Insert the long screw (Part "P") through the hole in the side of the plastic divider, Figure F11, and tighten into the right side frame of the ice maker mechanism.

12. Remove the insulation covering the ice maker thermostat. The green lead is used to ground the drive motor and the thermostat. (See balloon Figure F11). The double green lead spade connector will connect to the drive motor and the flag connector will connect to the thermostat. IT IS IMPERATIVE THAT THESE LEADS BE CONNECTED FOR PROPER GROUNDING BEFORE PROCEEDING WITH STEP 13. Reinstall the thermostat insulation and route the green lead out the slot provided. Insulation must be tight against the ice maker mechanism.

13. The electrical plug for the ice maker mechanism should be hanging down in front of the mechanism. Press the electrical plug onto the three small terminals protruding from the lower left-hand corner of the ice maker mechanism. The electrical plug will only engage in correct position, Figure F11. Position leads away from thermostat terminals.

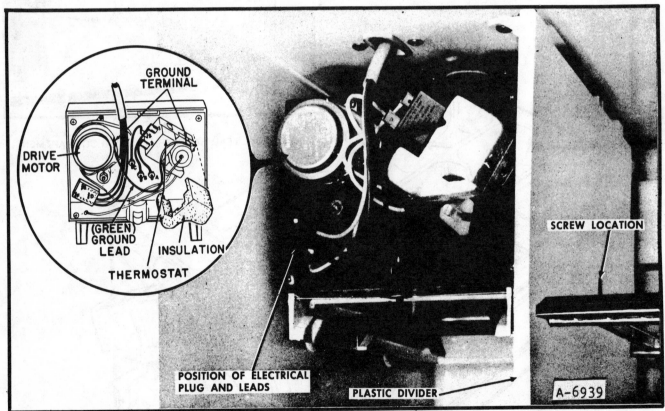

Figure F-11
ICE MAKER MECHANISM INSTALLED, MODEL FPC13-206TD

14. Mount the ice maker front cover (Part "F") to the ice maker mechanism. Position the tab (located at top rear of the cover) over the slot provided at the top of the ice making mechanism; swing the cover down and over the bottom of the ice maker mechanism while at the same time positioning the electrical lead harness in the groove provided at the top of the cover, Figure F12. Secure in place with the short flat head screw, (Part "R").

15. Press the control knob (Part "G") onto the shaft which extends through the ice maker front cover, Figure F12.

16. Slide the ice container (Part "B") in place, Figure F12.

17. Go to the rear of the refrigerator. Slide the rubber reducer cap of the electrical lead harness over the end of the $9/16$" O.D. electrical lead protector, Figure F13.

18.* Remove the machine compartment rear cover by removing the screws and push nuts. Save for reassembly in Step 35.

19. Install the lead covers (Parts "D" and "E") with screws (Part "N"), Figure F13.

20. Continue with the electrical lead harness installation by routing it down within the recess of the back of the cabinet, behind the black plastic lead covers, and into the machine compartment, Figure F13.

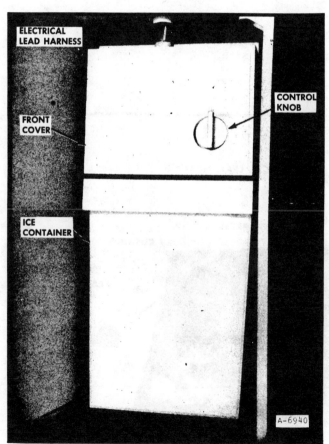

Figure F-12
ICE MAKER FRONT AND CONTAINER INSTALLED

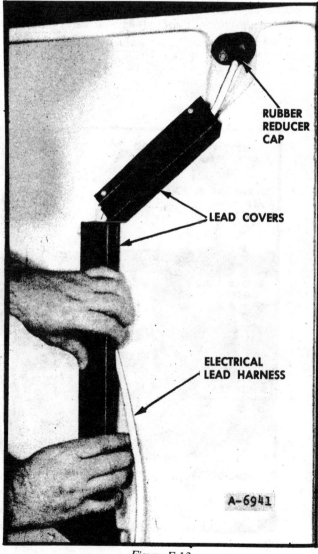

Figure F-13
INSTALLATION OF LEAD COVERS

21. Remove the fiber board baffle, Figure 15.

22. Remove the tape from the female connector block of the cabinet wiring. The block is hanging from the machine compartment top between the compressor and the condenser fan motor. Take the free end of the blue lead and insert into the end hole of the female block as shown, Figure 14. There are two tabs that snap into position inside the female connector block to prevent accidental removal.

23. Connect the male and female connector blocks together, Figure F14. They are polarized so that they will go together only one way.

24. Install the green ground lead to the condenser fan housing using the green screw (Part "O"), Figure F14.

25. Assemble the hose coupling (Part "Y"), supply tube adapter (Part "X") and the coupling washer (Part "W") finger tight to the water valve assembly (Part "V").

26. Assemble the plastic hex nut (Part "K") onto one end of the plastic fill tube (Part "J"). Press the red insert (Part "L") into this end of the fill tube and assemble to the water valve outlet. Turn the hex nut finger tight.

27. Install the remaining electrical connector block of the electrical lead harness to the water valve solenoid, Figure F14.

28.* Install the plastic protector bushing (Part "Z") into $^{13}/_{16}$" hole in the cabinet foot gusset, Figure F14.

29. Feed the plastic fill tube through this protector bushing from the inside out, Figure F14. Position the water valve behind the cabinet foot gusset by inserting the bracket through the slot from inside *while keeping the fill tube straight to prevent kinking,* Figure 15. Secure with a short pan head screw (Part "Q").

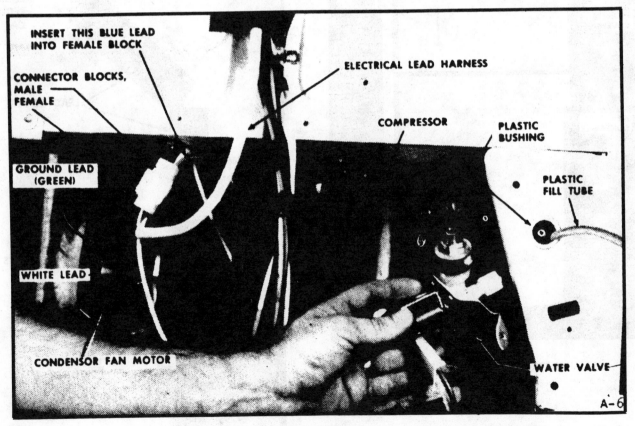

Figure F-14
INSTALLATION OF WATER VALVE

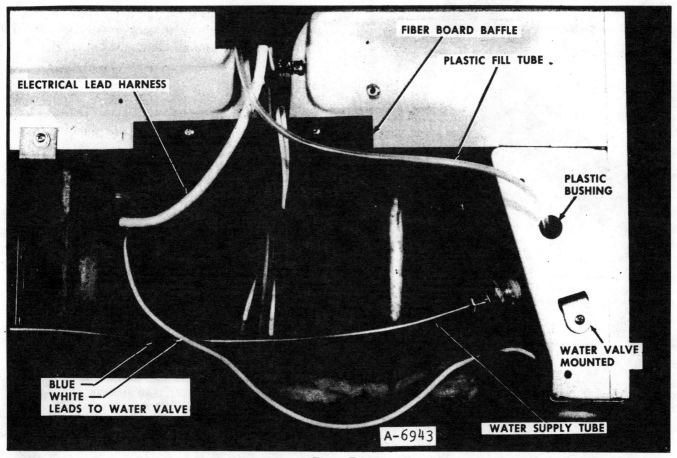

Figure F-15
WATER SUPPLY TUBE INSTALLATION MODEL

30.* Assemble clamp (Part "M") onto the upper end of the plastic fill tube and then install the fill tube onto the water tube connection at the top rear of the cabinet. Start the end of the fill tube onto the connection and "roll" it on as far as it will go. Position the clamp properly on the end of the tube, Figure F16.

31. Press the plastic fill tube into the groove of each black plastic lead cover, Figure F16.

32. Reinstall the fiber board baffle that was removed in Step 21, Figure F15.

33.* Flush any impurities from water supply line before connecting to refrigerator water valve. Install the nut (Part "U") and sleeve (Part "T") on the end of the copper supply line. Insert the supply line into the supply tube adapter (Part "X"), now on the valve. While holding the line all the way into the adapter, tighten with a wrench. Route the supply line to the left, Figure F15. Turn on the water supply and check complete system for leaks.

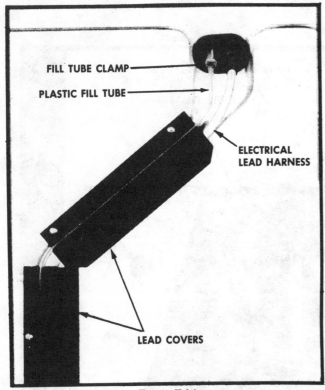

Figure F-16
FILL TUBE INSTALLATION

36. Plug in refrigerator service cord. Move refrigerator back into place and level product.

37. Place the Ice Maker Operating Instructions (Part "AA") into the product or present it personally to the user. The product is now ready for use.

For FPC13-220VU models, substitute the following steps when installing the IMK-4. The difference is mainly the routing of the electrical lead assembly and the plastic fill tube.

7a. Insert the electrical lead harness (Part "A") from the inside of the machine compartment through the hole in the rear cabinet foot gusset, Figure F18. Continue with Step 8.

18a. Remove the tape from the female connector block of the cabinet wiring located between the compressor and the cabinet side at the machine compartment top. Take the free end of the blue lead and insert into the end hole of the female block, Figure F18. (See another view of the connector blocks in Figure F14.) There are two tabs that snap into position inside the female connector block to prevent accidental removal. Continue with Step 23.

34.* Cut on the perforated lines of the machine compartment rear cover, Figure F17, being careful not to cut the fibrous insulation. While positioning the cover, fold out the tab near the top of the cover and route the plastic fill tube behind this tap and into the bottom holes, Figure F17.

35. Route the blue and white water valve leads over the tab and into the top holes, Figure F17, being careful not to kink the plastic fill tube. Remount the cover with the screws and push nuts removed in Step 18.

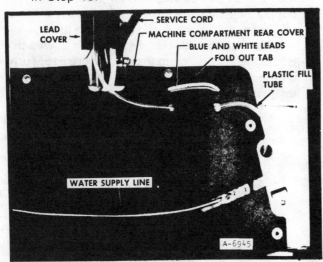

Figure F-17
MACHINE COMPARTMENT REAR COVER

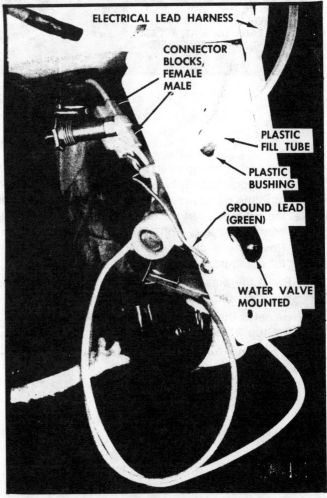

Figure F-18
WATER VALVE AND FILL TUBE

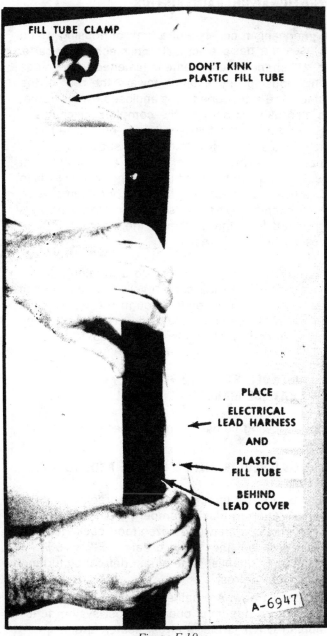

Figure F-19
INSTALLATION OF HARNESS AND FILL TUBE

33a. Because of a standardized electrical lead harness on this model there is excess blue and white leads from the valve. Push a loop of these leads through the plastic protector bushing from the inside, Figure F20. A small screwdriver may aid in pulling loop through grommet.

34a. Cut on the perforated lines of the machine compartment rear cover, Figure F21, being careful not to cut the fibrous insulation. Remount the cover with the screws and push nuts removed in Step 18. Fold out the tab at the cut perforations and route the plastic fill tube and the excess loop of blue and white leads behind this tab, Figure F21. Secure tab as shown. Finish with Steps 36 and 37.

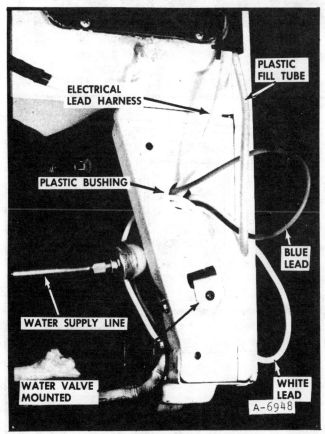

Figure F-20
WATER SUPPLY TUBE INSTALLATION

23a. Install the green ground lead to the side of the rear cabinet foot gusset using the green screw (Part "O"), Figure F18. Continue with Step 25.

28a. Feed the plastic fill tube through this protector bushing from the inside out, Figure F18. Position the water valve behind the cabinet foot gusset by inserting the bracket through the slot from inside *while keeping the fill tube straight to prevent kinking,* Figure F18. Secure with short pan head screw (Part "Q"). Continue with Step 30.

30a. Route the electrical lead harness and the plastic fill tube down the rear corner of the cabinet behind the black plastic side lead cover, Figure F19. Continue with Notes A and B.

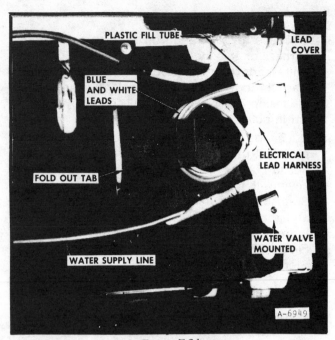

Figure F-21
MACHINE COMPARTMENT REAR COVER

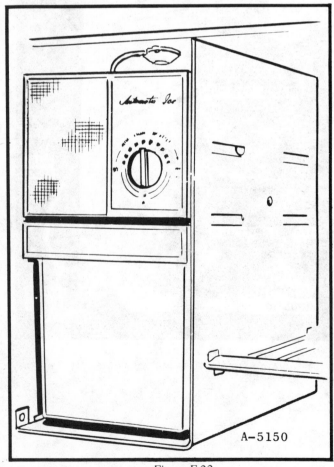

Figure F-22
IMK-1 "ADD-ON" AUTOMATIC ICE MAKER

OPERATIONAL CHARACTERISTICS

The mechanism consists of a control housing which encloses the basic electrical and mechanical parts. The "U" shaped metal frame is fastened to the sides of this housing and extends toward the rear of the cabinet. The eight cube tray is enclosed by this frame, it is also used to support the complete mechanism. This support has a full length slot on the left side which slides over two mounting supports in the freezer liner. Three support hooks are on the right side, two of these hooks fit into the horizontal slots in the vertical freezer barrier, the third hook remains on the mechanism. The ice server or storage bin is suspended from the bottom of the mechanism, and slides between two tracks.

The track assembly is coupled to a weighing device which controls the ice cube level. The shaft of the cube level control extends through the front of the housing cover. A control knob is pressed onto the cube level control shaft. See Figure F23.

RECOMMENDED STEPS TO FOLLOW ON A TRAY ICE MAKER SERVICE CALL

Check following items routinely at each service call.

ESSENTIAL REQUIREMENTS FOR ICE PRODUCTION:

1. Proper freezer temperature. Take freezer product temperature by placing temperature probe or thermometer in or between frozen food packages and record it on the work order. 0°F. to 5°F. is desirable. Establish time of last defrost by turning timer into defrost.
2. Proper air flow. Are baffles in place? Are air outlet and return openings open? Defrost system functioning O.K.? (Timer, defrost heater, drain.)
3. Ice level control between "ON" and "FULL". Note where it is set, record it on the work order. (10 increments)
4. Is ice container pushed in all the way? Is ice container free to move up and down, in and out? If motor starts when bucket is lifted, this indicates that the ice maker was shut off by the weigh switch. In this case, skip Step 5 and proceed to Step 6.c
5. Perform the manual thermostat reset procedure. This procedure is designed to get the ice maker in operation again. If water supply is interrupted while ice maker is in operation, or if thermostat does not reset for any other reason, turn ice level control knob to OFF or RESET. Wait 15 seconds and turn it back to the desired cube level setting.
6. Cabinet and ice maker must be properly leveled. Level the cabinet first, then the ice maker.

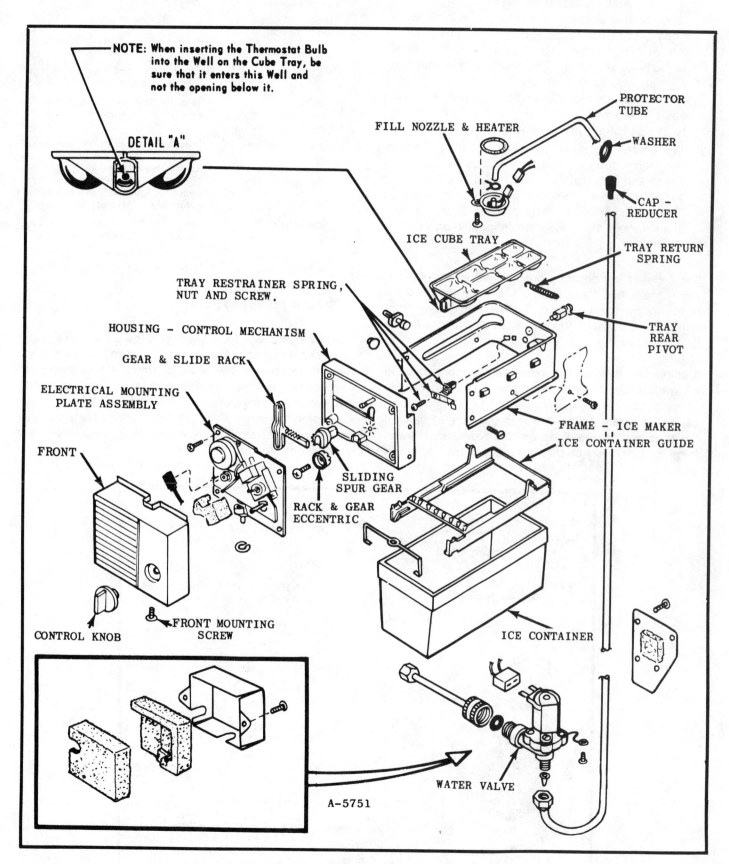

NOTE: When inserting the Thermostat Bulb into the Well on the Cube Tray, be sure that it enters this Well and not the opening below it.

DETAIL "A"

FILL NOZZLE & HEATER

PROTECTOR TUBE

WASHER

CAP - REDUCER

ICE CUBE TRAY

TRAY RETURN SPRING

TRAY RESTRAINER SPRING, NUT AND SCREW.

HOUSING - CONTROL MECHANISM

GEAR & SLIDE RACK

ELECTRICAL MOUNTING PLATE ASSEMBLY

TRAY REAR PIVOT

SLIDING SPUR GEAR

RACK & GEAR ECCENTRIC

FRAME - ICE MAKER

ICE CONTAINER GUIDE

FRONT

CONTROL KNOB

FRONT MOUNTING SCREW

ICE CONTAINER

WATER VALVE

A-5751

117

Figure F-23

7. Remove ice maker front.
   Is thermostat insulation in place?
   Is the bulb line formed correctly (stays close to the heater, does not touch electrical terminal)?
   Are bellows heater leads routed under bulb line?
   Is the bellows heater taped on the thermostat tight enough with aluminum tape?
   Is the electrical mounting plate cracked or damaged in any way?
   Is three-prong plug on properly?

8. The entire ice maker can be checked by shorting out thermostat terminals 2 and 3 so that you can watch the tray rotate, twist, snap the restrainer spring, return to horizontal and get a fill of water. Empty bucket must be in place or weigh switch shaft should be pulled down gently during this procedure.

If ice maker is operating normally when the service call is made and there is no evidence of the problem reported by the customer, follow the above eight points to check out the ice maker. (If it passes all these checks it is ready for ice production.)

If check No. 8 is not successful or if a more detailed analysis is needed for any other reason, follow the block diagram sequence on pages 12, 13 and 14. To use the diagram: follow arrow to the right of block if the statement is true. Follow arrow downward if statement in block is not true. Choose proper option if given.

DIAGNOSIS — NO ICE PRODUCTION

Possible Cause	Remedy
Not connected to a water supply line.	Connect to a water line.
Water line shut off.	Turn on water supply line valve.
Blocked water line.	Inspect the 80 mesh filter screen in the fill valve for foreign matter.
Momentary interruption of water supply.	Reset the mechanism as outlined in the "Proper Use of Ice Maker" section.
Freezer temperatures too warm.	Check freezer package temperatures. They should be plus 8°F. or colder. See "Testing Procedures."
Fill valve solenoid.	Check the fill valve solenoid operation as pointed out in "Testing Procedures."
Frozen fill tube.	Check the fill tube heater resistance as pointed out in "Testing Procedures."
Motor and Electrical Plate Assembly.	Check the Motor and Electrical Plate as pointed out in "Testing Procedures."

TESTING PROCEDURES
FREEZER TEMPERATURES

Freezer package temperatures should be plus 8°F. or colder. If the controls are set to the coldest positions resulting in a minus 6°F. to minus 10°F. package temperature, it is possible that the ice maker won't make ice. The bellows of the thermostat may have become the controlling point at this low temperature and will not permit the thermostat to reset when the water enters.

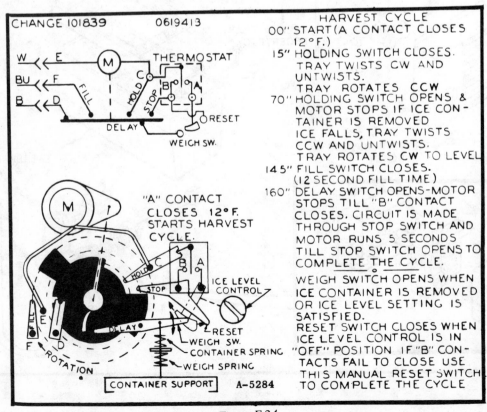

Figure F-24
ICE MAKER INFORMATION AND WIRING DIAGRAM LABEL

HIGH VOLTAGE OR LOW RESISTANCE THERMOSTAT HEATER — G. E. THERMOSTAT ONLY.

Remove front cover.

Line voltage of 130 volts or more requires a colder freezer temperature to compensate for added heat produced by the G. E. thermostat heater. A low resistance heater can also produce added heat.

POWER SUPPLY

Remove the front cover.

With a test lamp inserted in terminals "D" and "E" of the female ice maker connector plug, Figure F24,115 volts should be present when the refrigerator is connected to a power source.

WATER SUPPLY FILL VALVE SOLENOID

Remove the front cover.

With a jumper inserted between terminals "D" and "F," Figure F24, of the female ice maker connector plug, water should flow when the refrigerator is connected to a power source.

Note: If you get no water, reset the Ice Level Control.

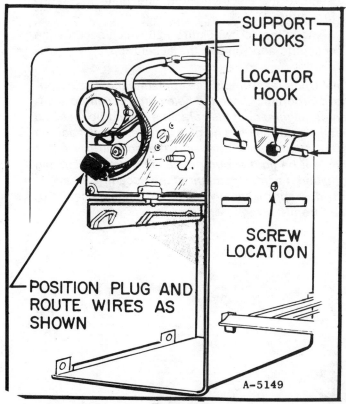

SUPPORT HOOKS

LOCATOR HOOK

SCREW LOCATION

POSITION PLUG AND ROUTE WIRES AS SHOWN

A-5149

Figure F-25
ICE MAKER MOUNTING DETAIL

MOTOR AND ELECTRICAL PLATE ASSEMBLY

Remove the front cover.

Place the empty Ice Server in position and turn the Ice Level Control Switch to the "OFF" position. Wait 15 seconds. Turn the Ice Level Control Switch to the "FULL" position. If the motor fails to run, connect a jumper to terminals "A" and "C". Terminals "A" and "C" are located adjacent to the thermostat and the "A" and "C" are formed into the mounting plate for identification. If the motor begins to run, the thermostat is not cold enough or it is inoperative. Recheck the freezer package temperature.
Caution: Use extreme care when working with jumper leads. The jumper should be fused (low amperage) and be certain proper terminals are used or a short circuit could result.

SERVICE OPERATIONS
A. GENERAL:

The following operations pertain to models equipped with the IMK-1 Automatic Ice Maker. These models provide for electrical grounding of the exterior doors, food liner, cold storage liner and water valve to the cabinet shell. Therefore, whenever an electrical grounding lead is removed during servicing, *be sure to reconnect it.* A grounding condition exists with a copper water line connected to the fill valve, as well as through the 3-wire service cord. *Be sure to disconnect the product from the source of power before attempting any service on the Automatic Ice Maker.*

A label, showing the sequence of the harvest cycle is located on the inside of the ice maker front. The numbers 00", 15", 70", to the left of the various stages of the harvest cycle represent seconds from the start of the cycle.

B. FRONT, ICE MAKER—REMOVAL:

1. Remove the container.
2. Remove the knob by pulling it straight off.
3. Remove one screw on the bottom of the front. Pull front out and up.

C. ASSEMBLY, ICE MAKER—REMOVAL:

1. Remove the Ice Maker Front, Procedure B.
2. Remove the lamp shield and lamp, if applicable.
3. Remove one screw on the side of the ice tray shelf frame, Figure F25.

4. Disconnect the ice maker mechanism 3-pronged electrical plug by pulling it straight off to avoid bending the 3 male prongs.
5. Lift up on the support hooks, Figure F25, and pull forward on the ice maker assembly for complete removal.

If the ice maker assembly is removed from a cold freezer compartment, it must be dry and at room temperature before replacement, to avoid the possibility of condensate forming on the commutator and electrical contacts within the electrical mounting plate assembly.

D. PLATE ASSEMBLY, ELECTRICAL MOUNTING—REMOVAL:

1. Remove the ice maker assembly, Procedure C.
2. Remove the "C" washer securing the guide support to the ice level control arm.
3. Remove four screws on front of this plate assembly and pull straight out.

Caution: Upon reassembly be sure the projection on the drive gear aligns with the groove of the gear and slide rack or damage to all drive parts will result.

Note: This assembly is the "heart" of the mechanism. Replacement as an assembly is recommended because of the precise adjustments required and because the cost of the assembly is less than individual parts plus the necessary labor to replace and adjust.

First production of this assembly will have an externally mounted G. E. thermostat partially covered with a piece of foamed insulation. The electrical leads from this thermostat are connected to the respective terminals on the front.

E. GUIDE, ICE CONTAINER—REMOVAL:
GUIDE SUPPORT

1. Remove the Ice Maker Assembly, Procedure C.
2. Remove the "C" washer which secures the guide support to the ice level control arm.
3. Spring out the tabs which secure the guide to the rear of the ice maker frame and remove the guide.

F. FRAME, ICE MAKER—REMOVAL:

1. Remove the Ice Maker Assembly, Procedure C.
2. Remove the guide, Procedure E.

3. Remove the two screws on each side of the frame.
4. Remove the frame by pulling off and tilting the ice tray so that the rear tray pivot can be pulled through the hole in the rear of the frame.

G. TRAY—REMOVAL:

1. Remove the frame, Procedure F.
2. Pull off the tray from the sliding spur gear.
3. Transfer the tray pivot to the replacement tray. When reinstalling the tray, align tray hole with bulb of thermostat.

H. PIVOT, TRAY—REMOVAL:

Remove the tray, Procedure G. The pivot is removed from the tray by using a screwdriver to push the pin out of the tray toward the rear.

I. SPRINGS—REMOVAL:

The tray restrainer spring and the tray return spring replacements are quite accessible when the ice maker assembly is removed, Procedure C.

J. ECCENTRIC, RACK AND GEAR—REMOVAL:

1. Remove the electrical mounting plate assembly, Procedure D.
2. Remove the screw, Figure F26, which secures the eccentric and remove the eccentric.

Note: Upon reassembly, the adjustment of this eccentric is important. Since the hole in the eccentric is off-center, alignment of the serations should be made so that the gear rack has no play between the flat portion of the sliding spur gear and the eccentric. Neither should it be tight enough to cause a drag or binding condition. Each adjustment of the eccentric serations provides approximately .001" movement against or away from the rack.

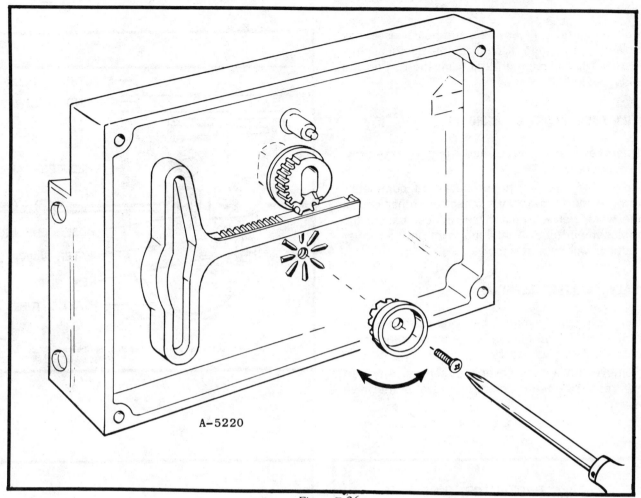

Figure F-26
RACK AND GEAR ECCENTRIC

K. RACK, GEAR AND SLIDE—REMOVAL:

Remove the rack and gear eccentric, Procedure J. The rack can be removed at the same time.

L. GEAR, SLIDING SPUR

1. Remove the rack and gear eccentric, Procedure J, and the gear and slide rack, Procedure K.
2. Remove the sliding spur gear from the end of the tray.

Caution: Upon reassembly, the flat portion of the sliding spur gear must be down against the rack. It is possible to install it 180° out of phase.

M. FILL TUBE AND HEATER ASSEMBLY— REPLACEMENT:

1. Remove the ice maker assembly, Procedure C.
2. Remove the screw that secures the fill tube and heater to the top of the freezer compartment.
3. Pull the reducer cap off of the protector tube and down on the plastic fill tube at the top rear of the cabinet shell. The plastic tube could become kinked upon reassembly if this is not done.

4. Pull the fill tube, Figure F27, and heater assembly down far enough to disconnect the electrical plug. Remove the plastic fill tube from the heater and plug assembly.

N. FILL TUBE, PLASTIC—REMOVAL:

1. Remove the fill tube and heater assembly, Procedure M.
2. Go to the rear of the product and completely remove the fill tube. Disconnect the other end at the water valve, transfer the reducer cap to the replacement tube. A little lubricant on the rubber reducer will help assembly.

O. VALVE, WATER—REMOVAL:

1. Depending on the model, the water valve is located at the top, Figure F28, or at the center rear of the cabinet shell, Figure F29.
2. Remove the access cover and screws. Any part can readily be replaced after removal of the valve assembly.
3. Be sure to transfer and reconnect the ground wire.

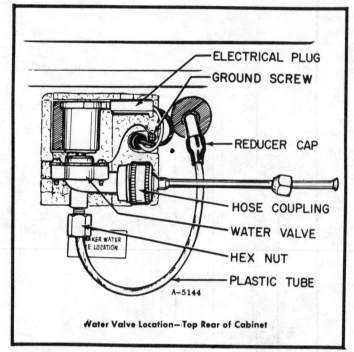

Figure F-28

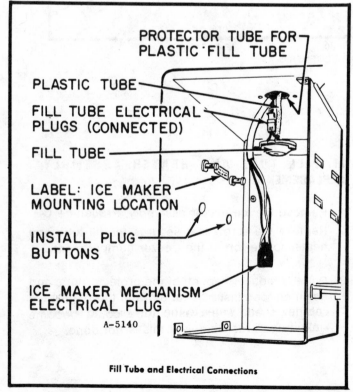

Figure F-27

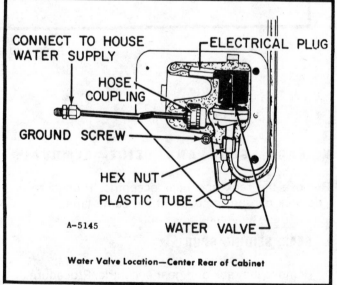

Figure F-29

P. CLEANING THE ICE MAKER:

The owner can clean the ice tray if desired, but . . . a few precautions must be observed. The ice tray is not easily accessible for cleaning. Under no circumstances should an attempt be made to clean the tray while it is in the horizontal position. There is always the possibility that the tray might be forced into the vertical position which could damage the mechanism.

When cleaning is considered necessary, empty the Ice Server and replace it, turn the Cube Level Control "OFF" and wait approximately 2 hours before attempting to clean the tray. At the end of 2 hours, turn the control to the "Full" position. Within approximately 2 minutes, the user can hear the ice cubes falling from the tray into the Ice Server. Then the refrigerator should immediately be disconnected from the electrical outlet. This will stop the ice maker with the tray in a partially inverted position. In this position, with the ice server removed, the tray can be easily cleaned.

The molds in the tray should be wiped out with a damp cloth and a solution consisting of 3 tablespoons of baking soda in one quart of warm water. Harsh abrasive cleansers or heavy duty liquid cleaners should never be used. After cleaning, the electrical cord should be plugged in and the control knob should be checked to be sure that it is in a "FULL" position.

In some areas the water supply may contain minerals, etc., that may form deposits in the cavities, preventing the cubes from being fully ejected. If, in the judgment of the serviceman, the tray needs cleaning, removal of the assembly from the refrigerator makes the job much easier.

Caution: If the ice maker assembly is removed, it must be dry and at room temperature before replacing because of the possibility of the formation of condensate on the commutator and electrical contacts within the electrical mounting plate assembly.

Q. SHAFT, WEIGH SWITCH—REMOVAL:

1. Remove the electrical mounting plate assembly.
2. Turn the Ice Level Control to OFF.
3. Remove the commutator gear.
4. Remove the spring contact blade which lays over the weigh switch shaft.

5. Reinstall "C" washer on the ice maker weigh switch shaft. This will insure that the two springs and washer will stay on the shaft. The spring with more and finer coils is on top, the heavier spring is on the bottom; a washer must go between them. The washer must be located so that the electrical mounting plate limits its movement.
6. The adjustment of the weigh switch is critical. Upon reassembly, proceed as outlined under Switch, Weigh Adjustment.

R. SWITCH, WEIGH — ADJUSTMENT. IMK-1 ONLY.

Note: Before adjusting the weigh switch, perform the following operation to assure that the weigh switch adjusting screw will retain its setting.

1. Remove the electrical mounting plate.
2. Loosen the weigh switch adjusting insert which is glued in place at the factory. Set it at the middle of the adjustment thread, Figure A and B.
3. Place a ⅜" long piece of foamed tape in the cavity where the weigh switch adjusting insert fits in the control mechanism housing, Figure E. This will insure that the insert will hold its position after adjustment. Newer models come with this tape already in place.
4. Reassemble the electrical mounting plate.
5. Reinstall ice maker in refrigerator. Leave front cover and thermostat insulation off and do not connect electrical plug for the time being.
6. Turn the ice level knob to the full clockwise position.
7. Connect an ohmmeter to the A and D terminals, Figure D, set meter to XI. A battery powered continuity light can also be used.
8. There should be continuity between A and D terminals with the empty ice container inserted all the way into the guide. There should be no continuity from A to D with the ice container removed.
9. Set the ice container on a level surface. The lip on the bottom front edge of the container should hang over the edge of the level surface to maintain container level. Add water to the container to a depth of 2½ inches, Figure C.
10. Insert the container to the stop position, then GENTLY move the container up and down; this will, of course, cause the A to D circuit to open and close. However, the circuit should be closed after the container is released. If it remains open after the container is released, the weigh switch setting would be too low. (Too few cubes.)

The weigh switch adjusting insert should be turned in a clockwise direction ¼ of a turn at a time until continuity between terminals A and D is indicated when the ice container is released.

11. Next place the container back on the level surface and increase the water depth to 3 inches. Reinsert the container to the stop position and again gently move the container up and down. There should be no continuity after the container is released. If there is continuity from A to D the weigh switch setting is too high. (Too many cubes.)

Turn the weigh switch adjusting insert ¼ of a turn counterclockwise until above requirements regarding the continuity are satisfied.

Note: The weigh switch adjusting insert can be turned with a small pocket screwdriver. To make this insert turn freely, loosen the four electrical mounting plate screws. The ice container must be removed during this adjustment.

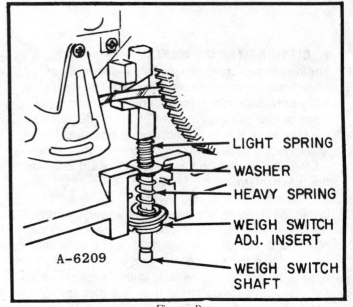

Figure B

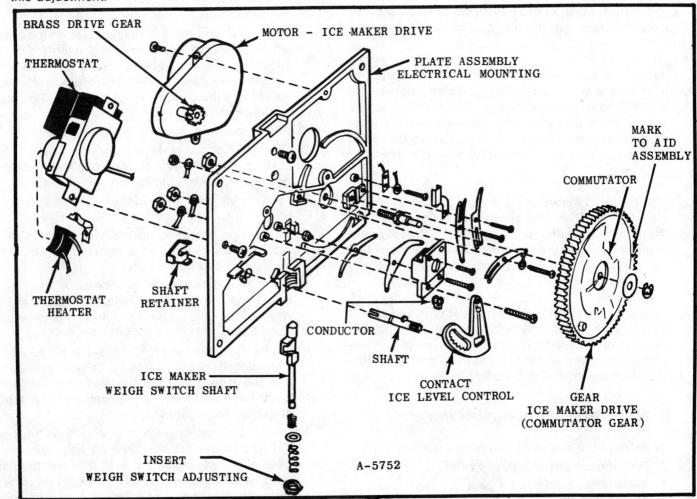

Figure A

Note: The 3 inch water level check is the maximum weigh switch setting. There may be some cube interference when the container is removed. Advise the user to reduce the cube level with the ice level control knob. Permanent damage to the internal components of the ice maker may result if the ice tray hangs up on the ice cubes in an overfilled ice bucket.

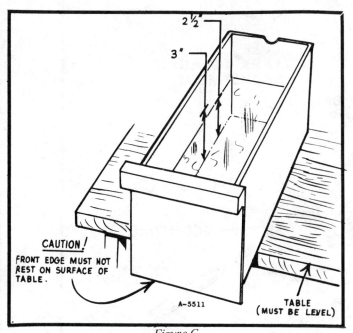

Figure C

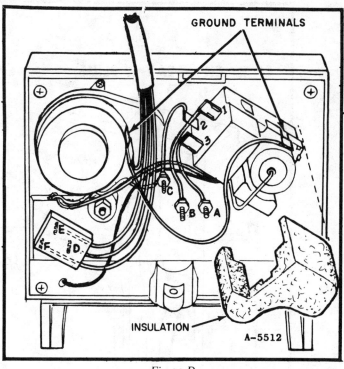

Figure D

S. ECCENTRIC, GEAR AND SLIDE RACK — REMOVAL AND ADJUSTMENT:

1. Remove the electrical mounting plate assembly.
2. Remove the screw which secures the eccentric. Remove the eccentric. The gear and slide rack can now be removed, Figure E.

Note: On reassembly, the adjustment of the eccentric is important. Since the hole in the eccentric is off center, alignment of the notches should be made so that the gear rack has no play between the flat portion of the sliding spur gear and the eccentric. Neither should it be tight enough to cause a drag or binding condition. The best way to achieve a neutral point is as follows.

The ice tray must be level. Hold the tray against the stop on frame at the right rear corner. Install the eccentric and tighten the screw. Check play and movement of gear rack. Loosen screw, adjust eccentric one notch tighter. Repeat above procedure until the gear rack binds. Adjust the eccentric once more by loosening it one notch.

A small mold mark is provided on the rim of the eccentric to serve as a guide during adjustment. See Figure E.

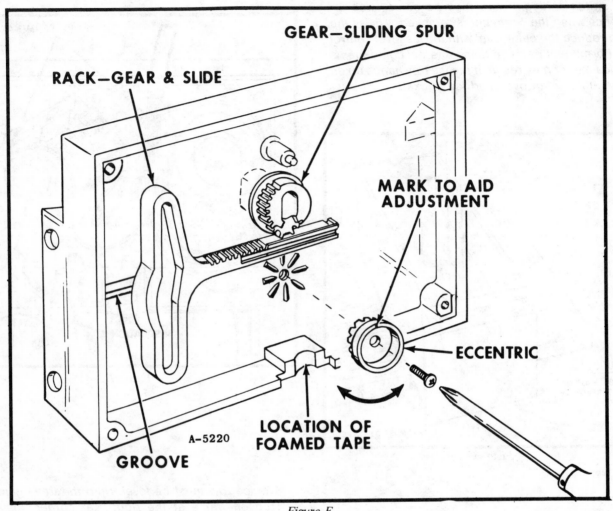

RACK—GEAR & SLIDE

GEAR—SLIDING SPUR

MARK TO AID ADJUSTMENT

ECCENTRIC

LOCATION OF FOAMED TAPE

GROOVE

A-5220

Figure E

DETAILED SUPPLEMENTARY OPERATIONAL INFORMATION

The tray type IMK-1 Automatic Ice Maker consists of these basic components and are identified in the figures which follow as:

1. Disconnect Plug (Points F, E and D)*.
2. Commutator.
3. Moveable Weigh Switch Arm*.
4. Moveable Ice Control Switch Arm* A (adjusted by user).

5. Fixed Holding Switch Arm* C.
6. Fixed Stop Switch Arm* B.
7. Thermostat.
8. Drive Motor.
9. Fill Valve.
10. Ice Container.

*The contacts on these arms are finger-type which bear against the commutator or another arm, Figure F30.

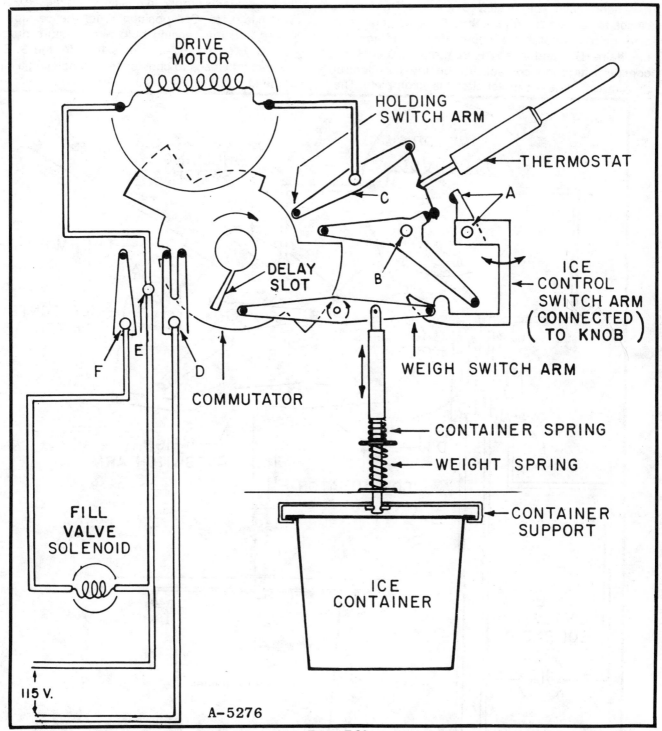

Figure F-30
FREEZING CUBES—HORIZONTAL TRAY POSITION

OPERATION OF ELECTRICAL AND MECHANICAL COMPONENTS

Figure F30 shows the motor and electrical mounting plate assembly in the freezing position after the tray has been filled with water. Since the Thermostat closes to contact "B" on the Stop Switch Arm at sensing bulb temperatures above plus 14°F. for the Frigidaire HT-2 and plus 17°F. for the G. E., there is no continuity through contact "A" on the Ice Control Switch Arm, thus the motor circuit is incomplete. The

commutator is attached to the large wheel, it is driven directly off of the motor gear; whenever the Motor runs, the Commutator must also rotate in a clockwise direction.

Ice Container is empty, in position, and the Ice Control Switch is set to "ON" position. When the ice cubes are frozen and the sensing bulb termperature drops to plus 12°F. for the HT-2 or plus 14°F. for the G. E. the thermostat closes to contact "A," completing a circuit to the Motor. This begins the harvest cycle.

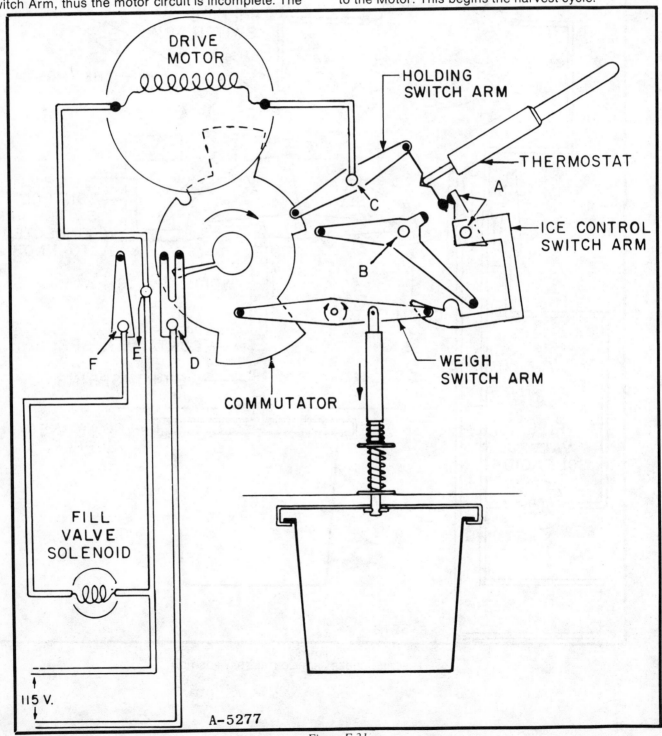

A–5277

Figure F-31

START OF HARVEST CYCLE—TRAY TURNS CLOCKWISE

HARVEST CYCLE

In Figure F31 the Commutator continues to advance until the holding switch (contact "C") on the Holding Switch Arm is closed, which completes the Motor circuit directly to the Commutator. It is at this point that the tray begins its first twist (clockwise). Notice that the circuit cannot be interrupted by the Weigh Switch or the Ice Control Switch until the first twist has been completed, and the tray has been returned to a normal horizontal position and on past to a position just before the cubes fall from the tray, Figure F32.

HOLDING SWITCH OPEN

The Holding Switch, Figure F32, opens when the tray is at approximately a 45° angle, to prevent continuation of the cycle and expulsion of the ice cubes in the event the Ice Server has been removed after the start of the harvest cycle. With the Ice Server in place, current to the motor continues to be supplied through contact "A" on the Ice Control Switch Arm and the Weigh Switch Arm contact.

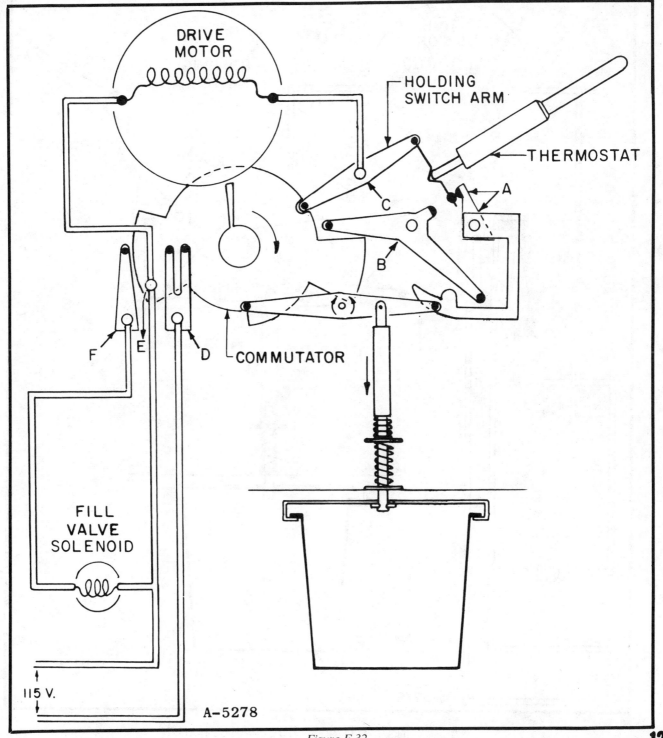

Figure F-32
HOLDING SWITCH OPEN—TRAY TURNED 45° COUNTERCLOCKWISE FROM HORIZONTAL

FILL

In Figure F33 contact "F" is now closed, completing the circuit from the Commutator to the Fill Valve. The

Holding Switch (contact "C") on the Holding Switch Arm is closed to prevent interruption of the Motor circuit during "FILL," thus producing a 12 second fill period.

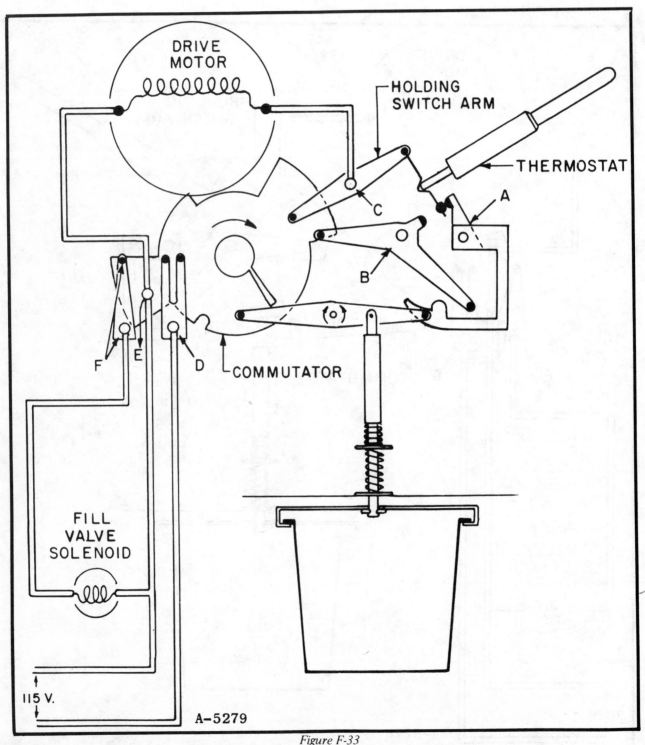

A–5279

Figure F-33

FILL CYCLE—HORIZONTAL TRAY POSITION

DELAY

In Figure F34 the water fill is now complete. The Holding Switch (contact "C") on the Holding Switch Arm is now open, thus the circuit may be complete through contact "A" on the Ice Control Switch Arm and on through the Weigh Switch Arm only until the contact on the commutator end of the Weigh Switch comes into the Delay Slot. This Delay Slot is an open, or dead spot, in the Commutator, which prevents continuous operation of the Motor in the event no water entered the tray. When the thermostat sensing bulb is warmed by the water, the Thermostat will close to contact "B" on the Stop Switch Arm completing the circuit and the Motor will advance. If water fails to warm the thermostat, the unit will have to be reset by turning the Ice Control Switch to "OFF," as shown in Figure F35, reestablishing a circuit to the motor.

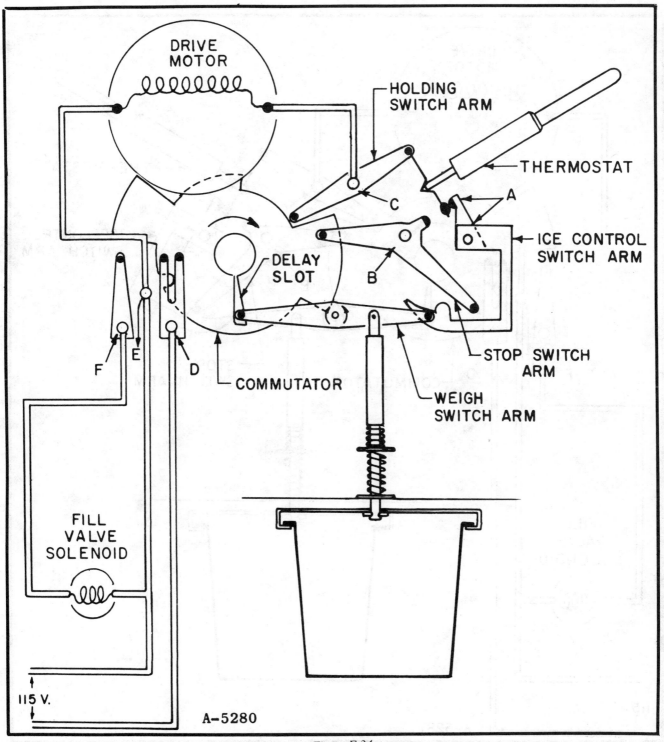

A-5280

Figure F-34
DELAY PERIOD—HORIZONTAL TRAY POSITION

MANUAL RESET

When the Ice Control Switch is turned to the "OFF" position, Figure F35, a circuit is completed through contacts "B" on the Stop Switch Arm and the Ice Control Switch Arm so the Motor may run until the Commutator reaches a normal completed cycle.

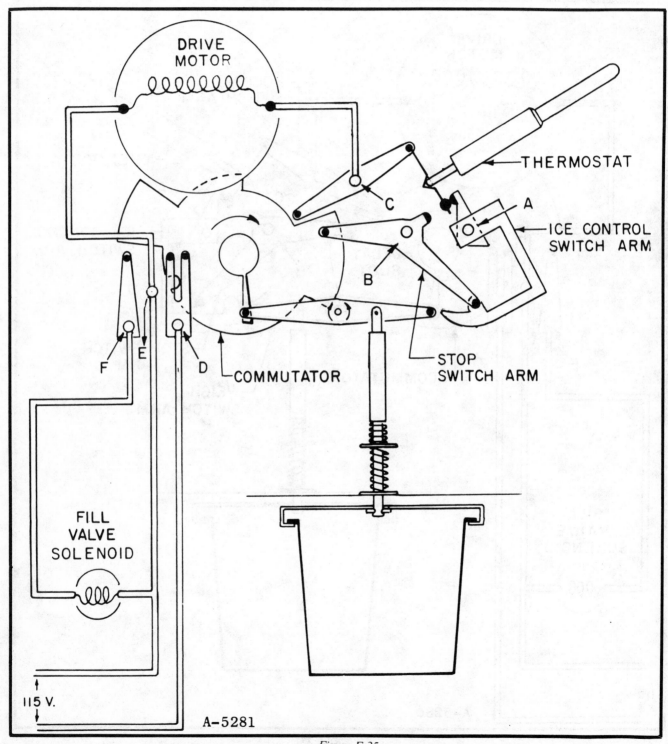

Figure F-35
MANUAL RESET—TRAY IN HORIZONTAL POSITION

OFF POSITION

In this position, Figure F36, the Weigh Switch is open. Contact blades "B" on the Stop Switch Arm and "C" on the Holding Switch Arm are both off of the Commutator, thus all circuits are open until the control is turned to "ON" position.

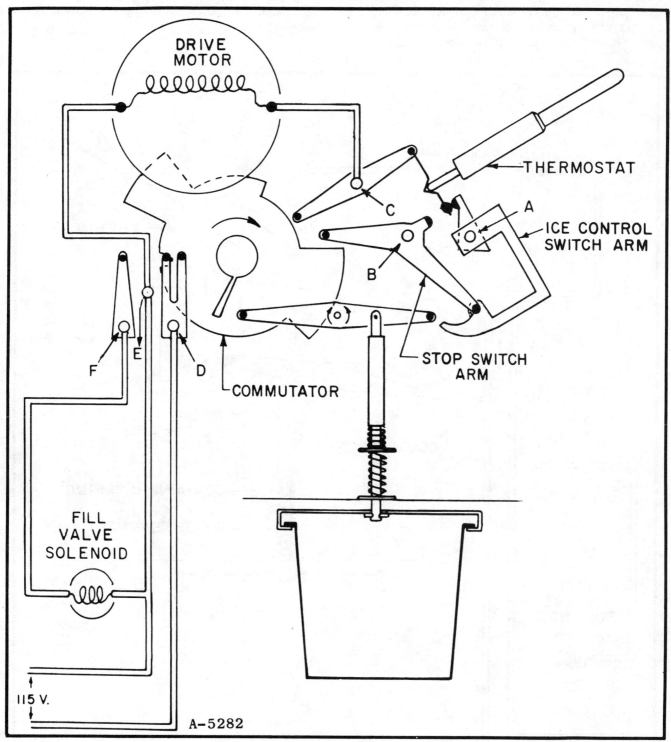

Figure F-36
OFF POSITION HORIZONTAL TRAY POSITION

ICE SERVER REMOVED

When the Ice Server is removed, Figure F37, the Container Spring expands, causing the shaft and Weigh Switch contact Arm to travel in an upward direction above the Ice Control Switch, opening the contact with the Ice Control Switch Arm. When the Ice Server is in place and full of ice cubes, the Container Spring and the Weigh Spring collapse, causing the shaft and Weigh Switch Arm contact to travel in a downward direction below the Ice Control Switch Arm, opening the contact. Since the Ice Control Switch Arm contact area is crescent shaped, the amount of ice produced can be varied by the adjustment of the Ice Control Switch.

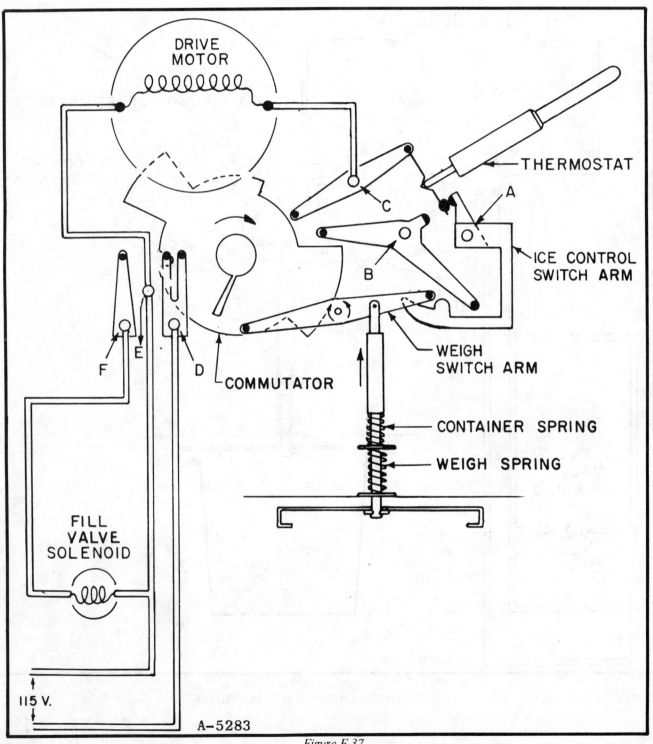

Figure F-37
ICE SERVER REMOVED

ICE MAKER MODELS IMK- ① ② ③ ④ ⑤ ⑥
IDENTIFICATION OF PARTS
(Figure F38)

This article is offered to show the differences in parts used in production, and parts used by service, and to point out the variations in parts used on the different Ice Maker models. Parts are sometimes duplicated to avoid confusion.

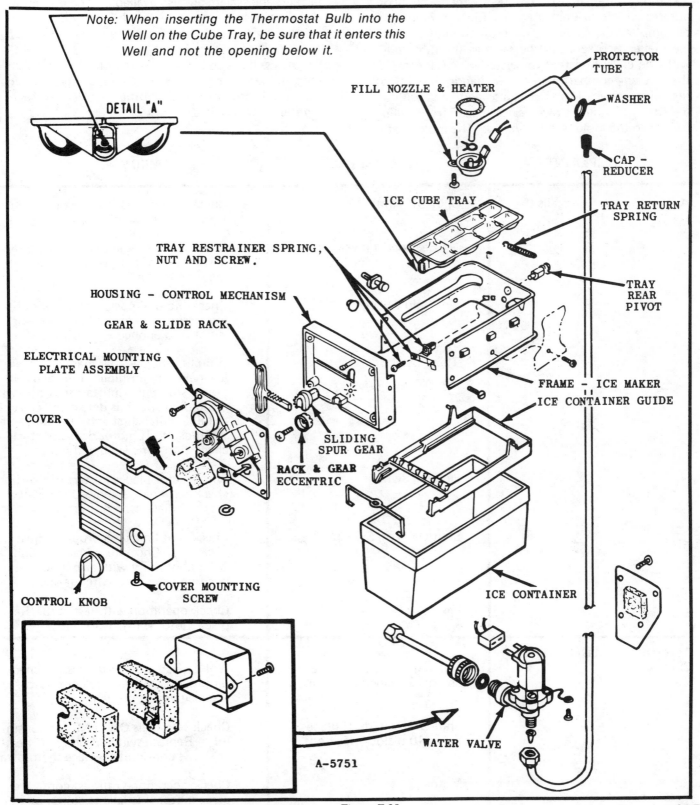

Note: When inserting the Thermostat Bulb into the Well on the Cube Tray, be sure that it enters this Well and not the opening below it.

DETAIL "A"

FILL NOZZLE & HEATER

PROTECTOR TUBE

WASHER

CAP – REDUCER

ICE CUBE TRAY

TRAY RETURN SPRING

TRAY RESTRAINER SPRING, NUT AND SCREW.

HOUSING – CONTROL MECHANISM

GEAR & SLIDE RACK

ELECTRICAL MOUNTING PLATE ASSEMBLY

TRAY REAR PIVOT

COVER

SLIDING SPUR GEAR

RACK & GEAR ECCENTRIC

FRAME – ICE MAKER

ICE CONTAINER GUIDE

CONTROL KNOB

COVER MOUNTING SCREW

ICE CONTAINER

WATER VALVE

A-5751

Figure F-38

135

MODEL IK-1 (Refer to Whirlpool Design Page 24 for technical data).

Model IK-1 Ice Maker currently used in some models of Frigidaire refrigerators differ in many respects to the IMK series. The IMK Ice Makers are dependent upon the mold heater and the flexing of the mold to accomplish the harvest. The IK-1 which also incorporates a mold heater uses an Ice Ejector, *Figure A6* to complete the harvest. The installation of the IK-1 Ice Maker is practically the same as the IMK series with the exception that the water valve is located in the machine compartment rather than in the back of the refrigerator. The IK-1 is also equipped with an Ice Stripper which prevents ice pieces from falling back into the mold, see *Figure A4*. The mold heater

Figure A3 is rated at 165 watts and covered with an aluminum sheath. The heater assembly is imbedded in a groove provided on the underside of the ice mold. The mold heater is wired in series with the ice maker thermostat, which also acts as a safety device.

The original heater can be replaced if necessary. The fastners and sealant are also available. The operational characteristics are outlined under Whirlpool Design starting on page 24 and continuing through page 37.

Diagnosis charts and wiring diagrams to aid in troubleshooting can be found on the following pages, included is an illustrated parts identification list. Refer to *Figures F-39 and F-40.*

COMPLAINT	POSSIBLE CAUSE	REMEDY
Automatic Icemaker fails to start	Shut-Off Arm	Check that arm is in lowest position Lower arm if raised.
	No Electrical Power	Check for power at black and white ice maker leads. Correct cabinet wiring if defective.
	Not Cold Enough	Check mold temperature at a mounting screw. If above 10°, evaporator is is not cold enough.
	Thermostat	If mold is below 10°, manually start ice maker by pushing timing gear. If motor starts, thermostat, shut-off switch or holding switch is defective. Check holding switch first and, if O.K., replace thermostat. If motor does not start, check shut-off switch and motor.
	Holding Switch	With ejector blades in starting position, check terminals C and NC for continuity. Replace switch if open.
	Shut-Off Arm Switch	Check that linkage is proper, adjust if necessary. Check terminals NO and C for continuity with arm in lowest position. Replace switch if open.
	Motor	Check operation with test cord. Replace motor if it fails to start.
Automatic Icemaker fails to Complete cycle	Holding Switch. If blades are in 10 o'clock position.	With switch plunger depressed, check terminals C and NO for continuity. Replace switch if open.
	Shut-Off Switch. If blades are in 10 o'clock position.	Check terminals C and NC for continuity. Replace switch if open. If heater shows continuity replace thermostat.
	Motor	Check operation with test cord. Replace motor if it fails to start.

COMPLAINT	POSSIBLE CAUSE	REMEDY
Automatic Icemaker fails to stop at end of cycle	Holding Switch	With ejector blades in starting position, check terminals C and NO for continuity. Replace switch if closed.
Automatic Icemaker continues to eject when bin is full	Shut-Off Arm	Check that linkage is proper. Switch should open when arm is in raised position. Adjust if required. Check terminals C and NO for continuity with arm raised. Replace switch if closed.
Automatic Icemaker produces undersize Ice pieces.	Mold	Check for level. Adjust if required.
	Water Supply	Check that supply line and water valve strainer are completely open and that adequate pressure is maintained. Clear restrictions or advise customers accordingly.
	Water Valve Switch	Test-cycle IM and measure water fill. Adjust switch if required.
	Thermostat Short-Cycling indicated by ice shells or hollow ice in storage bin.	Check thermostat bond to mold. Assure good thermal contact with alumilastic. Check thermostat calibration by replacing with new part.
Automatic Icemaker spills water from mold	Mold	Check for level. Adjust if required. Check top edge for evidence of siphoning. Prevent capillary action using silicone grease in this area.
	Water Inlet Tube	Check that inlet tube and fill trough fit properly and water does not leak during fill cycle. Adjust fit if required.
	Water Valve	Check that water does not enter mold after cycle is completed. Replace valve if leaking and water pressure is proper.
	Water Valve Switch	Test-cycle IM and check that water fill does not exceed volume capacity of mold. Adjust switch if required. With ejector blades in starting position, check terminals C and NC. Replace switch if closed.
	IM fails to stop at end of cycle	Refer to complaint, "Fails to stop at end of cycle."
	Thermostat short-cycling	Refer to complaint, "Undersize ice pieces."

COMPLAINT	POSSIBLE CAUSE	REMEDY
Water fails to enter mold during cycle.	Water Supply	Check that water line, valve, and valve strainer are open. Remove restrictions, open valve, or instruct customers accordingly.
	Water Valve	Observe inlet tube and fill trough for ice. If obstructed with ice, check water valve for slow leak. If valve leaks, check for proper water line pressure. Replace heater if open. Replace water valve if pressure is within specifications.
	Valve Solenoid Coil	Check terminals for continuity. Replace coil if open or shorted.
	Water Valve Switch	With plunger out, check terminals C and NC for continuity. Replace switch if open.

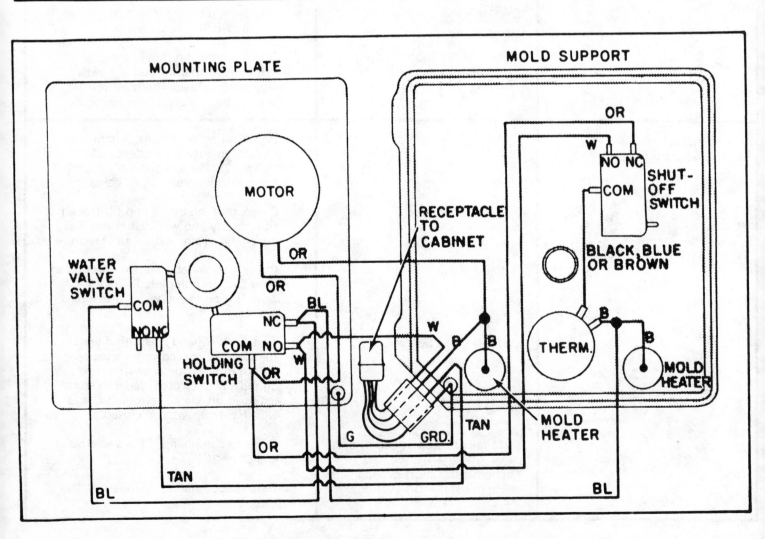

FIGURE F-39 **WIRING DIAGRAM FOR IK-1 ICE MAKER**

ICEMAKER PARTS

ORDER PARTS BY MODEL NUMBER

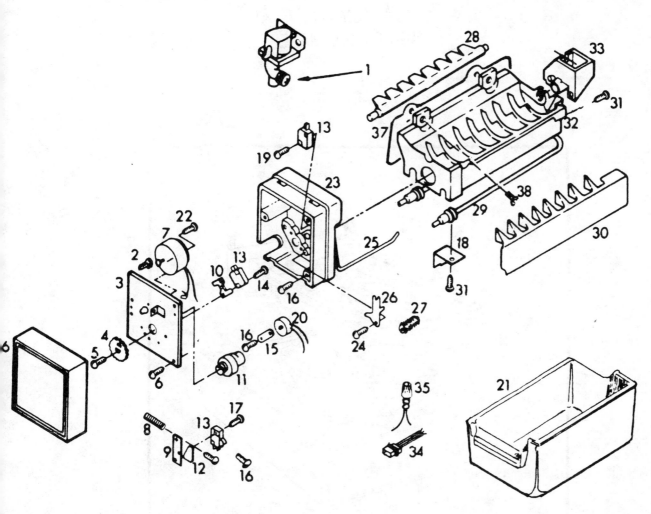

INDEX NO.	DESCRIPTION	INDEX NO.	DESCRIPTION	INDEX NO.	DESCRIPTION
1	Valve	15	Clamp	28	Ejector
2	Screw	16	Screw 8-32 X 1	29	Heater — mold
3	Plate — mounting	17	Screw 4-40 X 3/4	30	Stripper — ice
4	Gear	18	Bracket	31	Screw 8-32 X 3/8
5	Screw	19	Screw 4-24 X 3/4	32	Mould and heater assembly
6	Screw 8-18 X 1/2	20	Thermostat		(includes illus. 29)
7	Motor	21	Container — ice	33	Bearing and inlet
8	Spring		Trim — container	34	Wire assembly
9	Plate — valve switch	22	Screw — ground	35	Connector — wire
10	Spacer — holding switch	23	Support	36	Cover
11	Cam	24	Screw — shoulder	37	Shield
12	Insulator	25	Arm — shut off	38	Eyelete
13	Switch (3)	26	Arm lever		
14	Screw 4 - 40 X 1	27	Spring — shut off arm		

FIGURE F-40 **ILLUSTRATED PARTS IDENTIFICATION LIST**

TYPE

AUTOMATIC
ADD-ON ICE MAKER

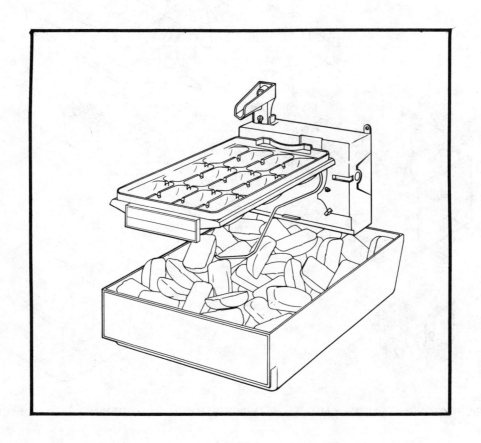

AUTOMATIC ICE MAKER

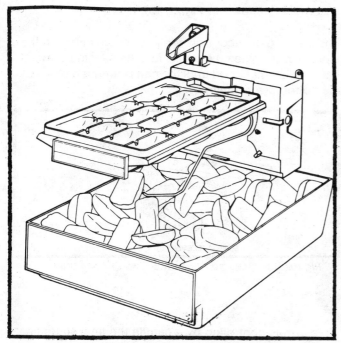

Figure D1—Ice Maker & Storage Tray

This Automatic Ice Maker, *Figure D1* operates on a timed cycle and will harvest at regular intervals if the temperature of the freezer compartment is below +15°F. A sensor wire senses the level of ice cubes in the storage bin and stops the harvest when the bin is full. The cycle can also be terminated manually by placing the sensor wire in the OFF position.

The water valve, located in the machine compartment, is a solenoid valve which, when energized, allows water to flow to the ice mold. It features a built-in constant flow device that allows the correct amount of water to enter the ice mold, as long as the water pressure is maintained between 12 and 120 pounds.

HOW ICE MAKER WORKS

To simplify the description of the ice maker operation, we have divided the complete cycle into two (2) sections, "Freeze Time" and "Harvest Time". Here's how it works:

As soon as the freezer compartment temperature descends to approximately +15°F., a thermostat, acting on a lever arm, energizes the timer motor switch and starts the motor. The timing gear is thus slowly turned through a high reduction gear train. The ice maker is now operating in the FREEZE TIME part of the cycle.

Near the end of the timing gear revolution, the wire sensor arm moves down into the cube storage bin. This is the beginning of the HARVEST TIME part of the cycle which lasts approximately 8 minutes. If the storage bin is not full and the sensor arm is permitted to continue unobstructed, it will return to its original position and the cycle will continue.

At this time, the cube tray begins to rotate. After a few degrees of rotation, the motor control switch is "locked in" and the HARVEST is allowed to continue, regardless of any type of shut-off (manual or through motor control thermostat).

After rotating approximately 180 degrees, one corner of the tray engages a stop which prevents that part of the tray from rotating further. As the shaft continues to turn, the tray is twisted into a severe bind which loosens the cubes from the tray surface. The tray stop is then retracted and the tray is rapidly released from the bind, thus shaking all ice cubes into the storage bin.

The tray continues to turn until it has completed a full 360 degree rotation. At this time, the fill switch is energized and the tray pockets are filled with the correct measure of water.

The ice maker is now ready to start a new cycle.

TIME PER CYCLE

This automatic ice maker has been designed to operate at three different freeze time speeds by changing the gear ratio inside the housing. The unit is shipped from the factory set on the normal cycle speed and only under unusual conditions should the cycle be changed.

After initial installation, or after the cabinet has been inoperative for an extended period of time, it is possible to experience a time lapse of more than 3½ hours before the first water fill. This is normal and is dependent upon the position in the cycle the ice maker was left in previously.

The length of time it takes to harvest the first ice cubes will depend on how much time is left on the ice maker clock motor plus the normal cycle time.

It is possible to initiate a manual harvest cycle which will be discussed elsewhere in this section.

141

TIME PER CYCLE

SHIFT LEVER POSITION	FREEZE TIME CYCLE	HARVEST TIME CYCLE
+ Time	206 Min.	8 Min.
Normal Time	146 Min.	8 Min.
- Time	106 Min.	8 Min.

If an unusual amount of door openings are encountered on a freezer compartment of moderate temperature, to insure solid cubes, it may be necessary to increase the time required for the ice maker to complete its cycle. If, however, the freezer chest temperature is extremely low with below average rate of door openings, it may be desirable to reduce the cycle time so more ice cubes will be produced.

In order to change position of the gear shift lever, you must spring the lever away from the cover plate slightly to disengage it from the slots. If a gear shift selection is being made while the ice maker is mounted in the cabinet, you may have to use a small screwdriver to spring the lever back while making the selection.

After a gear selection is made, you should allow the motor to run for a few minutes and then check the lever to make sure it has engaged the slot for which the selection was made. It is always desirable to make a gear shift selection while the motor is operating because the gears mesh much easier while they are running.

ICE MAKER COMPONENTS
ICE TRAY

The ice tray is of molded polyethylene, with 12 separate molds. The tray has two metal inserts for added strength during the shucking operation. The tray is put into a severe bind and quickly released to shuck the ice cubes free (see *Figure D2*).

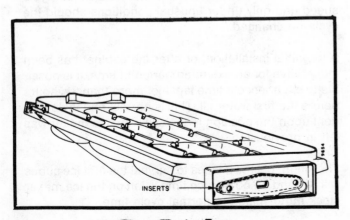

Figure D2—Ice Tray

SENSING ELEMENT

The sensing element is a wax type that actuates a lever arm to control the motor switch. This control will open the switch when the temperature rises above +19˚. The reset temperature is +15˚. The complete assembly is shown in *Figure D3*.

The sensing arm controls the level of ice cubes in the storage pan. When the pan is full, the sensing arm will shut the ice maker off until the pan is emptied or some ice cubes are removed.

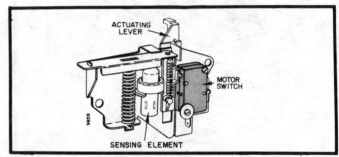

Figure D3—Sensing Element & Switch Assembly

This arm is controlled by a cam in the final stage timing gear. Sensing occurs just prior to engaging the harvest gear train (tray rotation).

This arm can be placed and locked in the "off" position for manual termination (see *Figure D4*).

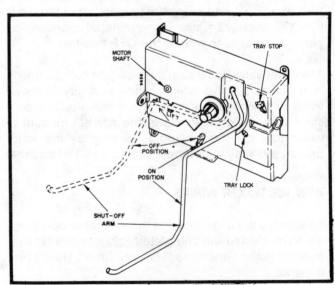

Figure D4—Sensing Arm Function

DRIVE GEAR TRAIN

The drive gear train is made up of several Delrin gears. The first stage timing gear is driven directly off the motor gear which in turn drives the gear shift assembly as shown in *Figure G5.* The final stage timing gear is driven by the pinion shaft of the shift assembly. It has a molded in cam groove that actuates the cam slide for the sensor arm. This gear also actuates the tray lock to release the tray and, at the same time, energizes a lug on the harvest gear. As the final timing gear continues to rotate, it turns the harvest gear only a few degrees until its teeth mesh with those of the first stage timing gear. The harvest gear is now turned quite rapidly through one complete revolution to its original position. Since the harvest gear has a flat spot adjacent to the first stage timing gear, it will remain stationary until actuated by the second stage timing gear or by releasing the tray lock and rotating the tray slightly.

GEAR SELECTOR

The ice maker is provided with a gear selector that will vary the time required for the ice maker to complete its timed cycle (see *Figure D5).* When the ice maker is shipped from the factory, the gear selector is set on the normal position (center). Only under unusual conditions should the cycle be changed.

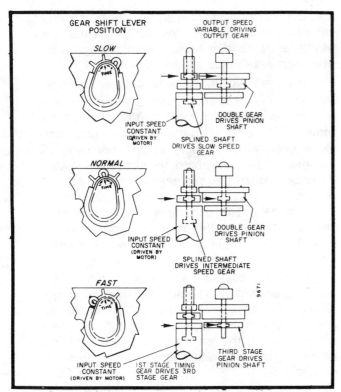

Figure D5—Gear Shift Lever & Gear Function

SWITCHES

The two switches used are single-pole, single-throw type. They are not interchangeable for service. Their main functions are:

MOTOR CONTROL SWITCH (normally closed) is connected in series with the timer motor. It is only opened by the actuator for the sensor arm during termination, or by the thermostat and lever arm if the freezer compartment is too warm. The actuator prevents the switch from opening during the harvest sequence to assure motor operation until completion of the tray fill.

WATER VALVE SWITCH opens the water valve during the fill cycle. This is the only adjustable component of the ice maker.

WATER FILL ADJUSTMENT SCREW

This screw is the only component adjustment that can be made on this ice maker. The screw is calibrated so that one complete turn is equal to about 20 cc's. of water. Turn the screw clockwise to decrease it.

FILL TROUGH

This part is molded Delrin and is used to support the inlet tube and direct the water into the ice tray.

DRIVE MOTOR

A low wattage permanent magnet type motor is used. It has a single output shaft on which a Delrin gear is mounted. The shaft turns one revolution per minute (1 RPM) (see *Figure D6).*

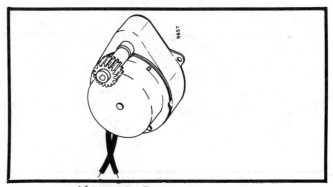

Figure D6—Timing Motor

CHECKING ICE MAKER

It may be necessary on occasion to test cycle an ice maker to check its operation. This can be done on the repair bench or while mounted in the cabinet. (Special test cord shown in *Figure D7*).

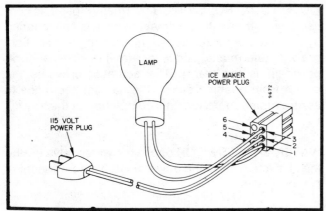

Figure D7—Special Ice Maker Test Cord

A special ice maker test cord (part #RA43123) can be made for bench repair using the cabinet ice maker connector and the remote valve wiring harness (see *Figure D7*).

To manually start a cycle, first place the sensor wire in the sensing or ON position. Pull the motor control actuator lever, near the bottom of the housing, down and block it (see *Figure D8*). You can determine if the ice maker motor is operating by observing the motor shaft on the front of the housing.

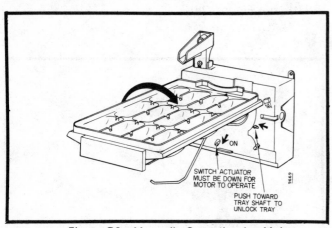

Figure D8—Manually Operating Ice Maker

CHECKING WATER VALVE

A quick check of the water valve circuit can be made as follows:

1. Disconnect the lead wires from the water valve and attach a volt meter or test lamp to them.

To manually cycle the ice tray push the lever on the right (see *Figure C8)* toward the tray to release the tray lock. Turn the tray clockwise until the gears are engaged. When the harvest cycle is complete the lever will return to its normal position.

> NOTE: Initiating a manual harvest will place the next "normal" harvest, out of phase in the timing sequence. The possiblity of the next harvest occurring before the cubes are frozen and dumping water or ice shells into the storage pan exists. The tray should always be emptied after a manual harvest to prevent water being dumped into the storage pan.

2. Run the ice maker through a manual harvest cycle. Toward the end of the cycle, voltage should be indicated at the valve leads for a period of approximately 13 seconds indicating that the switch and wiring are OK. If no indication of voltage is present, the trouble is in the switch or wiring. If voltage is present but the valve does not operate when connected in the circuit, the valve is defective.

3. If voltage is not present in step 2, check the water valve switch and wiring. You can check the open wiring for damage or breaks. If no damage is evident, remove the ice maker from the cabinet.

4. Remove the cover from the mechanism case.

5. Using a small screwdriver, push the fill switch lever toward the switch. A click should be heard. If not, remove the switch from the ice maker and check for a click while pushing in and releasing the plunger. If no click is heard, the switch is defective. If a click is heard, you can determine if the contacts are operating by placing ohmmeter clips across the terminals. The ohmmeter should indicate a closed circuit when the plunger is depressed. If a closed circuit is not indicated, the switch is defective.

6. If the switch checks OK, a mechanical malfunction could be caused by (1) a bent fill switch lever arm; (2) defective drive gear cam lobe; (3) improper adjustment of switch. These functions can be observed by running the ice maker through a manual harvest cycle using a test cord like the one shown in *Figure D7*.

CHECKING SENSING ELEMENT

If the temperature of the freezer chest is stabilized at +5° or 10°F. or below, the sensing element should be reset (sensing element turns switch on at approximately +15°F.). You can determine if it is reset by pushing down on the motor switch actuator shown in *Figure D8.* If the freezer chest is cold enough and the sensing element is operating properly, you should experience only a very short travel of the actuator and the click of the micro switch will not be heard.

> *NOTE: Make sure the sensing arm is in the operating position and unobstructed while performing this test. Otherwise the motor switch would be held in by the sensor arm and the click of the micro switch would not be heard. Therefore, a service man may be led to believe the sensing element is operating properly when actually it may not be.*

If it is determined that the temperature of the freezer is low enough to reset the sensing element but you can push the switch actuator down and hear the micro switch click, you can assume the sensing element is defective.

CHECKING WIRING

Most of the wiring and connections inside the ice maker can be inspected by removing the rear cover from the ice maker case. *Figures D9* and *D10* illustrate the ice maker wiring in pictorial and schematic form.

CHECKING GEAR TRAIN

A visual inspection of the gear train will usually reveal any faulty gears or cams. You can reverse the procedure outlined in *Figures D12* through *D18* to dismantle the ice maker.

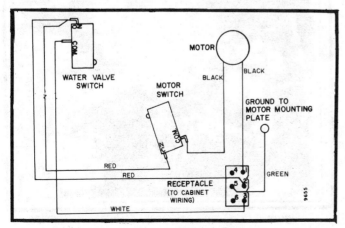

Figure D9—Pictorial Wiring Diagram

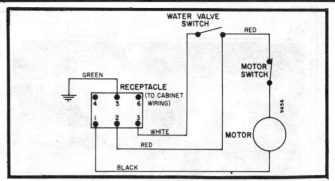

Figure D10—Schematic Wiring Diagram

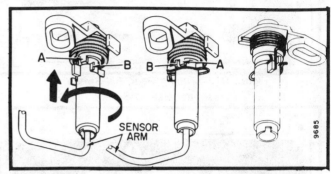

Figure D11—Sensor Arm Hub & Lever Assembly

To assemble the sensor arm hub and lever, insert the sensor arm in the hub as shown in *Figure D11.* Place the spring and lever on the hub and rotate the hub in direction of the arrows until tab "B" passes beyond tab "A". Now push the components together and install the retainer clip. You can compare the assembly with the illustrations in *Figure D11* to determine if it has been properly assembled. You can now remove the sensor arm.

CHECKING MOTOR SWITCH

This is a normally closed switch. If the ice maker motor fails to run, make sure the sensor arm is not OFF. Check the actuator for the control switch on the front of the ice maker to make sure it is down. **Remember the temperature of the sensing element must be below +15°F.** for the switch to close. If the actuator is not down, you can push it down with your finger then check to see if the motor shaft is turning (see *Figure D4*). If the motor does not run under the above conditions, remove the ice maker from the cabinet. Then refer to step 1 on page 6.

1. Using a test cord like the one shown in *Figure D7*, connect power to the ice maker.
2. Place the sensor arm in the operating position and, at the same time, push the switch actuator down. The switch plunger should be clear of the actuator and sensor arm levers. If it is not, there is a mechanical malfunction that prevents the switch from closing.

Figure D12—Timing Gear, Fill Switch
Adjustment Gear & Screw

Figure D13—Fill Switch Lever

Figure D14—Fill Switch, Harvest Gear Assembly
including Spring

3. If the plunger is free but the motor still does not operate, carefully short across the switch terminals. If the motor runs, the switch is defective. If it does not run, check the motor and wiring for defects.

CHECKING MOTOR

Place the sensor arm in the operating position. If the motor switch actuator is not down, hold it down with your finger while watching the motor shaft at the upper left side of the housing. If the motor does not turn, check the ice maker power plug. If this does not reveal the problem, remove the ice maker from the cabinet and proceed as follows:

1. Remove the cover plate from the mechanism case.

2. Remove the motor mounting plate — it is held in position by three screws (see *Figure D18*). Do not remove the motor from the plate nor disconnect any wires.

3. Apply power to the ice maker using a test cord such as the one shown in *Figure D7*.

4. A voltage check to determine if power is being supplied to the motor can be made as follows: Push one probe of the voltmeter in along side the motor lead wire until it touches pin 1 of the connector block. Touch the other probe to the common terminal of the motor switch while holding down on the switch actuator on the front of the case. If voltage is present, but the motor does not run, the trouble can be locked gears or defective motor.

5. Remove the motor mounting plate — held in position by three screws (see *Figure D18*). Do not remove the motor from the mounting plate yet.

6. Energize the motor again by holding the switch actuator down. If the motor shaft turns, the trouble could be in the gear train. If it does not turn, the motor is defective and must be replaced.

Figure D15—Tray Lock Assy. & Retainer, Sensing
Element & Motor Control Switch Actuator, Wire Harness

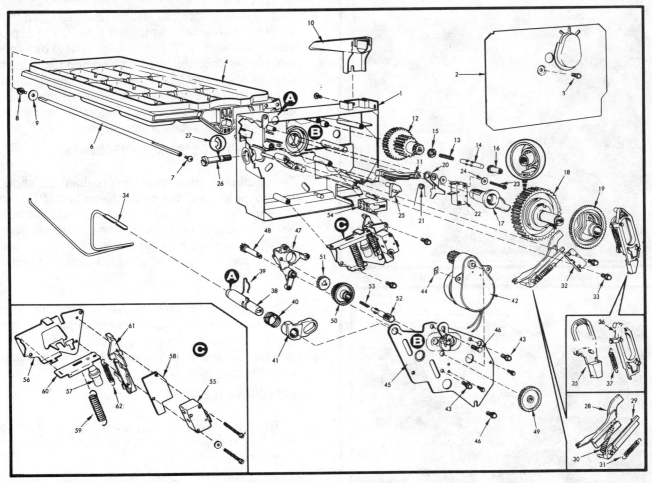

Exploded View of Ice Maker

Automatic ice maker (complete)
Nylon nut, ice maker mtg.
Mechanism assembly
1 Case, mechanism
2 Cover, mechanism case
3 Screw, cover mtg.
4 Ice tray (less trim plate)
6 Drive shaft, ice tray
7 Screw, drive shaft mtg. to harvest gear
8 Screw, drive shaft mtg. to tray
9 Washer, drive shaft mtg. to tray
10 Fill spout — Fill tube
   Clamp for fill tube
11 Tray stop
12 First stage timing gear
13 Spring, first stage gear train
14 Spline shaft, first stage gear train
15 Intermediate speed gear
16 Slow speed gear
17 Tray, drive hub
18 Harvest gear assembly (includes spring)
19 Final stage timing gear
20 Fill switch lever
21 Spring, fill switch lever
22 Water fill switch (normally open)
23 Screw, fill switch mtg.
24 Washer, fill switch mtg.
25 Fill switch adjustment gear
26 Screw, adjusting fill switch
27 Retainer, fill switch adjustment screw
28 Tray lock
29 Release, tray lock

30 Spring, tray lock
31 Spring, lock release
32 Retainer, tray lock
33 Screw tray lock mtg.
34 Sensor Arm
35 Actuator, sensor arm
36 Cam slide, sensor arm
37 Spring, cam slide
38 Hub, sensor arm
39 Retainer clip, sensor arm
40 Spring, sensor arm hub
41 Lever, sensor arm hub
42 Motor (includes drive gear)
43 Screw, motor mtg.
44 Nut, motor mtg.
45 Motor mtg. plate (includes selector lever)
46 Screw, motor plate mtg.
47 Support, gear shift assy.
48 Pinion shaft, shift assy.
49 Output gear, shift assy.
50 Double gear, shift assy.
51 Third stage gear, shift assy.
52 Output spline shaft, shift assy.
53 Spring, shift assy.
54 Wire harness & connector block
55 Motor control switch (normally closed)
56 Plate, motor control assy. mtg.
57 Temperature sensing element
58 Spacer, switch mtg.
59 Element return spring
60 Lever, sensing element
61 Actuator, motor control switch
62 Spring, actuator

Figure D16—Final Stage Timing Gear

Figure D17—Cam Slide, Actuator Hub & Lever for Sensor Arm

Figure D18—Motor Mounting Plate, including Motor & Gear Shift Assy.

148 **DISMANTLING & ASSEMBLING**

Figures D12 through *D18* illustrated the sequence in

which the ice maker mechanism should be assembled. This sequence can also be reversed to aid the service man in dismantling the ice maker. The exploded view on page 147 is also beneficial in determining the proper location of small components such as springs and small gears that are not clearly shown in this sequence of illustrations. Some of the individual assemblies in the exploded illustration are enlarged to give a more detailed view of them.

REPLACING ICE MAKER COMPONENTS

The exploded illustration is provided to show the relative position of the components and to familiarize the servicemen with the nomenclature used. In addition to the usual tools and equipment, the service man should have wire nuts for electrical splicing, silicone grease for use where specified and a special ice maker test cord.

If your customer complains that ice cubes are sticking in the tray, there is probably an abundance of mineral deposit in your community which is leaving a film residue on the tray surfaces. Have your customer soak the tray in vinegar.

REPLACING ICE TRAY

1. Remove the shield from in front of the ice maker by pulling the lower left front out until it pulls free of the retainer hooks.

   To replace this shield, press the top retainer loop into the upper hooks and "snap" the left side into place.

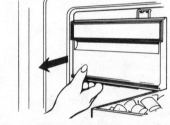

2. Remove the screw and washer from the front of the tray shaft.
3. Pull the tray straight off the shaft.
4. Install the replacement tray in the reverse order of removal.

REPLACING SENSOR ARM

1. Remove the shield from in front of the ice maker by pulling the lower left front out until it pulls free of the retainer hooks.
2. Remove the screw and washer from the front of the tray shaft and remove the tray.
3. Using a screwdriver, pry the sensor arm out of the mechanism case enough so you can remove it with your fingers.
4. Install the replacement sensor wire in the reverse order or removal.

REPLACING WATER VALVE SWITCH

1. Remove the ice maker from the cabinet.
2. Remove the cover from the mechanism case.
3. Disconnect the wires from the water valve switch.
4. Remove the switch mounting screw and lift the switch off the adjustment gear pin.
5. Install the new switch in reverse order of removal and install the rear cover.
6. Install the ice maker in the cabinet and check the water fill adjustment as outlined on page 3 of this section.

REPLACING WATER VALVE ADJUSTMENT SCREW

1. Remove the ice maker from the cabinet.
2. Remove the cover from the mechanism case.
3. Remove the split washer from the adjustment screw by pulling it straight out with needle nose pliers.
4. Remove the adjustment screw by turning it counterclockwise.
5. Install the new adjustment screw in reverse order of removal.
6. Install the ice maker in the cabinet and check the water fill adjustment as outlined on page 3 of this section.

REPLACING MOTOR CONTROL SWITCH

1. Remove the ice maker from the cabinet.
2. Remove the cover from the mechanism case.
3. Remove the motor mounting plate— it is held in place by three screws as shown in *Figure D18*.
4. Disconnect the lead wires from the switch terminals.
5. Remove the switch mounting screws and remove the switch.
6. Install the replacement switch in reverse order of removal and connect the wires.
7. Install the motor mounting plate. Care must be taken when doing this to make sure the harvest gear shaft, sensor arm hub is seated in the motor mounting plate and that the hub of the motor gear is seated in the opening on the front of the mechanism case. These components must all be aligned before the motor mounting plate screws are tightened. Normally, you can position the plate and rock it back and forth with your thumbs to align the components. Sometimes however, you may have to position a shaft with a screwdriver in order to get them aligned.
8. Install the mechanism cover plate.
9. Install the ice maker in the cabinet and check for proper operation.

REPLACING MOTOR

1. Remove the ice maker from the cabinet.
2. Remove the cover from the mechanism case.
3. Remove the motor mounting plate — it is held in place by three screws as shown in *Figure D18*.
4. Remove the two motor mounting screws (small head).
5. Disconnect the motor lead wire from the switch.
6. Cut the remaining motor lead off close to the motor. Strip about ½ inch of insulation from this lead.
7. Using a wire nut, connect the wire of the replacement motor to the lead that was prepared in step 6.
8. Connect the terminal lead of the replacement motor to the motor switch.
9. Install the motor on its mounting plate.
10. Position the wires between the motor and the motor switch and install the motor mounting plate. Care must be taken when doing this to make sure the harvest gear shaft and sensor hub are seated in the motor mounting plate and that the hub of the motor gear is seated in the opening on the front of the mechanism case. These components must all be aligned before the motor mounting plate screws are tightened. Normally, you can position the plate and rock it back and forth with your thumbs to align the components. However, sometimes you may have to position a shaft with a screwdriver in order to get them aligned.
11. Install the mechanism cover plate.
12. Install the ice maker in the cabinet and check for proper operation.

REPLACING SENSING ELEMENT

1. Remove the ice maker from the cabinet.
2. Remove the mechanism case cover plate.
3. Using needle nose pliers to hold the lever up, remove the sensing element as shown in *Figure D20*.
4. Install the replacement sensing element in reverse

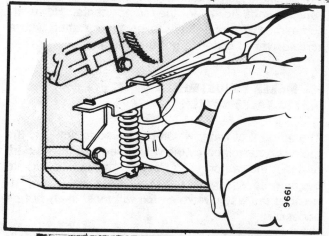

Figure D20—Replacing Sensing

149

order of removal. Make sure the sensing element plunger is seated in the indentation of the lever.

5. Install the mechanism case cover plate.

6. Install the ice maker in the cabinet and check for proper operation.

REPLACING SENSING ELEMENT RETURN SPRING

1. Remove the ice maker from the cabinet.

2. Remove the mechanism case cover plate.

3. Using needle nose pliers to hold the lever up, remove the sensing element as shown in *Figure D20*.

4. Using pliers, pull the sensing element lever out enough to clear the actuator for the motor control switch. This will allow the lever to drop down so the spring can be removed or installed without tension on the spring.

5. Install the replacement spring and lift the lever arm up so it rests on top of the actuator for the motor control switch.

6. Install the sensing element in reverse order of removal.

7. Assemble the ice maker and install it in the cabinet — check for proper operation.

REPLACING GEARS & OTHER ASSOCIATED PARTS

Figures D12 through *D18* are a sequence of illustrations that can be used to help the service man install the gears and associated components in their proper location. When installing the final stage timing gear, the process is aided by holding the tray lock back as you set the timing gear in place. If the assembly sequence is followed correctly, there should be no difficulty encountered.

When you install the motor mounting plate, care must be taken to make sure the harvest gear shaft and sensor arm hub are seated in the motor mounting plate. Also make sure the hub of the motor gear is seated in the opening on the front of the mechanism case. The components must all be aligned before the motor mounting plate screws are tightened. Normally, you can position the plate and rock it back and forth with your thumbs to align the components. However, sometimes you may have to position a shaft with a screwdriver in order to get them aligned.

ICE MAKER ADJUSTMENTS
WATER VALVE SWITCH

The amount of water which is allowed to flow into the mold is determined by the valve used and the amount of time the water valve is energized. If it should become necessary to replace a water valve, adjustment of the water valve switch will most likely not be necessary.

To check the water fill, place a cup or other container below the inlet tube or water valve outlet. Manually cycle the ice maker and collect the water (if air is in the water line, throw away the first few samples). The manual cycle will take 8 minutes.

Measure the water sample from the cycle in a medicine bottle or baby bottle or any vessel graduated in cc's. You should receive 200 cc's + 10% from one fill cycle.

A polyethylene plug is used to cover the hole where the water valve adjustment screw is located. If adjustment of the screw becomes necessary, the plug can be removed with a sharp knife point or similar object.

The fill cycle can also be adjusted on a timed basis if an accurate means of measuring time is available. However, there is a slight variation in the fill time of a warm ice maker and a cold one. Therefore, the chart is provided so the service man can adjust the water fill time while servicing the ice maker on the bench.

FILL CYCLE DURATION

ROOM TEMPERATURE	FREEZER TEMPERATURE
13-1/4 seconds ±1/2 second	12-1/2 seconds ±1/2 second
DO NOT EXCEED 14-1/2 seconds	DO NOT EXCEED 14 seconds

Determine whether you want more or less water. The adjusting screw is calibrated so that one complete revolution changes the water fill about 20 cc's. Turning the screw clockwise decreases the fill while counterclockwise increases the fill (see *Figure D21*).

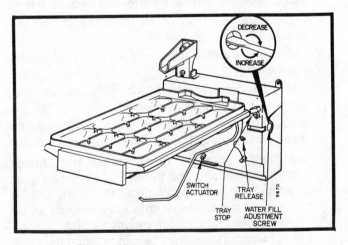

Figure D21—Water Fill Adjustment

Silicone grease, part number 43282, should be used any place where moving parts are involved. Ordinary grease will not allow the ice maker components to function properly under low temperature conditions. Therefore, it should not be used.

SERVICE HINTS

1. When operation of the appliance is to be discontinued for any length of time, the ice cube tray should be emptied and dried. This will prevent mineral deposits from forming on the tray.

2. The ice maker water valve is equipped with a built-in water strainer. If the installation is on sandy well water or if local water conditions require periodic cleaning or replacement, a second water strainer should also be installed in the ¼" water line.

   IMPORTANT: Before you initiate a manual harvest cycle, empty the ice tray and make sure the sensor arm is in the ON position.

3. When you manually run the ice maker through a cycle, the next normal cycle will be out of phase with the timing sequence and water or ice shells could be dumped into the storage pan.

4. If the ice pieces stick in the tray, the problem is usually caused by mineral deposits on the tray surface. The tray pockets can be cleaned with a cloth soaked in vinegar. If, however, the deposit is extremely heavy, it may be necessary to replace the tray.

Because the sensor arm operates off a cam on the final stage timing gear, you may or may not notice movement of the arm during a manual harvest cycle. This is because the final stage timing gear is not affected by a manual harvest cycle.

ICE MAKER TROUBLE SHOOTING

TROUBLE	POSSIBLE CAUSE	CORRECTION
ICE MAKER DOES NOT RUN	NOT COLD ENOUGH	Check to see if motor switch actuator is down. If it is not, check the freezer temperature. Temperature must be below 15° for motor to start.
	SENSOR ARM IN OFF POSITION OR STORAGE FULL OF ICE	Check position of sensor arm and level of storage tray.
	SENSING ELEMENT DEFECTIVE	Replacing sensing element.
	MOTOR CONTROL SWITCH DEFECTIVE	Replace motor control switch.
	DEFECTIVE WIRING	Check wiring and connector plug.
ICE CUBES CLUSTERED IN STORAGE TRAY	TOO HIGH FREEZER TEMPERATURE AFTER ICE CUBE HARVEST	Turn freezer control to colder position.
	SENSING ELEMENT DEFECTIVE	Replace sensing element.
	BROKEN SENSING ELEMENT RETURN SPRING	Replace spring.
	BROKEN SPRING ON MOTOR SWITCH ACTUATOR	Replace spring.

ICE MAKER SPILLS WATER FROM MOLD	MOLD	Check the mold for level. Level the cabinet if required.
	WATER INLET TUBE	Check that the inlet tube and fill trough fit properly and water does not leak during fill cycle. Adjust them if necessary.
	WATER VALVE	Check that the water does not enter the mold after the cycle is completed. Check the valve to see if it is leaking. If the valve is leaking and the water pressure is known to be correct, replace the valve.
	WATER VALVE SWITCH	Test cycle the ice maker and check that the water fill does not exceed the volume capacity of the mold. Adjust the switch if required.
WATER FAILS TO ENTER ICE TRAY	WATER SUPPLY	Check that water line valve and valve strainer are open. Remove any restrictions, open valve, or instruct customer accordingly.
	WATER VALVE	Observe the inlet tube and the fill trough for ice. If obstructed with ice, check the water valve for slow leak. If valve leaks, check for proper water line pressure. Replace the water valve if pressure is within specifications.
	WATER VALVE SOLENOID COIL	Check the terminals for continuity. Replace the coil if it is open or shorted.
	WATER VALVE SWITCH	Check the water valve switch as outlined on page 10.
	WIRING	Check the wiring for open circuits. Check the connector plug for defective or corroded terminals.
ICE MAKER CONTINUES TO EJECT WHEN BIN IS FULL	ACTUATOR FOR SENSOR ARM	Check to make sure the pin of the switch actuator is riding in the groove of the final stage timing gear.
ICE MAKER PRODUCES UNDERSIZED ICE PIECES	MOLD	Check the mold for level. Level cabinet.
	WATER SUPPLY	Check that the water line and water valve strainer are completely open and that adequate pressure is maintained. Clear the restrictions or advise the customer accordingly.
	WATER VALVE SWITCH	Test cycle the ice maker and measure the water fill. Adjust the switch if required.

WHITE-WESTINGHOUSE

The White-Westinghouse Ice Maker Unit is used on most of the White Consolidated Brands such as the Admiral, Franklin Brands Gibson, Kelvinator and many others, if you have any doubts as to what text to follow, look for identifying features, such as the shape of the ice cubes, the switches and gears (remove cover plate). This unit is also used on some models of General Electric and Hotpoint. Use the illustrations in the Ice Maker Repair Master for identifying the ice maker you are attempting to repair. All add-on Ice Makers should be removed from the refrigerator for service to avoid shutting down the refrigerator and for safety sake.

OPERATIONAL DESCRIPTION

COMPONENTS

ICE MOLD, Figure WW-1

This is a die cast aluminum part, with separators that mold 12 ice crescents. The unit is designed with a new front surface which completes the mold. The thermostat is bonded to the face of this surface.

A film of silicone grease is on the top edge of the mold to prevent capillary action from siphoning water.

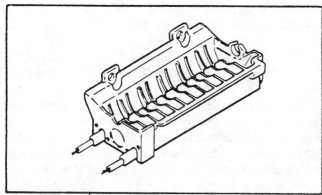

Figure WW-1, Ice Mold, Makes 12 Crescent Shaped Molds

MOLD HEATER, Figure WW-2

The mold heater is rated at 165 watts to thaw the ice free from the mold. The heater is in series with the thermostat which also acts as a safety device.

The production heater is staked in place on the bottom of the mold, for the first time this heater can be serviced separate from the mold. This is done by using four flathead retaining screws adjacent to the heater in the mold.

Alumilastic is used between the heater and mold to assure good thermal transfer.

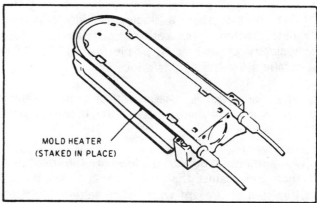

Figure WW-2, Mold Heater, replaceable   Staked in place

ICE STRIPPER, Figure WW-3

This is a new plastic part attached to the dumping side of the mold. It serves as a decorative side cover and the fingers are used to prevent ice pieces from falling back into mold.

ICE EJECTOR, Figure WW-4

The ejector is molded of delrin. Its twelve blades sweep the ice from the mold cavities during the ejection cycle.

The drive end of the ejector is "D"-shaped to assure positive one-way coupling. Both ends of this part are lubricated with silicone grease in the area of the bearings.

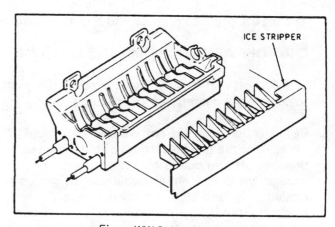

Figure WW-3, Ice Stripper

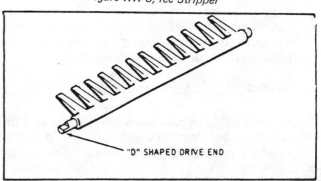

Figure WW-4, Ice Ejector (12 Blades)

THERMOSTAT, Figure WW-5

The thermostat is a single-pole, single-throw, bi-metal switch. This control starts an ice ejection cycle by closing at 18° plus/minus 5°. The reset temperature is 50° plus/minus 5°.

The thermostat is in series with the mold heater and acts as a safety against overheating in case of mechanical failure.

An alumilastic bond is made where the thermostat is mounted against the mold. A seal gasket is used in this area to prevent water from leaking into the support housing.

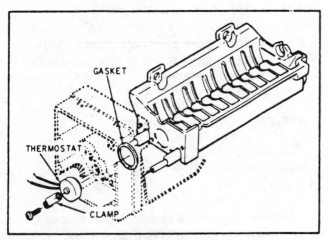

Figure WW-5, Thermostat and Gasket

SHUT-OFF ARM AND LINKAGE, Figure WW-6

The sut-off arm is cam-driven and operates a switch

SHUT-OFF ARM AND LINKAGE, Figure WW-6

The shut-off arm is cam-driven and operates a switch to control the quantity of ice produced.

Each ejection cycle the arm is raised and lowered during each of the two revolutions of the timing cam. If the shut-off arm comes to rest on top of ice in the storage bin during either revolution the switch will remain open and stop the ice maker at the end of that revolution. When sufficient ice is removed to lower the arm production will resume.

A manual shut-off is built into the linkage. By raising the arm to its highest extreme the lever arm will lock at that position until forced downward.

TIMING SWITCHES, Figures WW-7, 8

The three switches used are single-pole, double-throw type. They are identical except for function, and can be used interchangeably for service. Their main functions are:

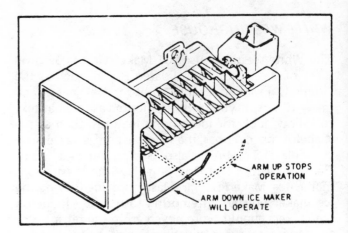

Figure WW-6, Shut-Off Arm

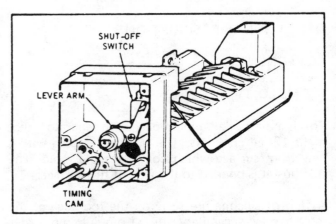

Figure WW-7, Shut-Off Switch

HOLDING SWITCH—assures completion of a revolution once the ice maker has started.

WATER VALVE SWITCH—opens the water valve during the fill cycle. This is the only adjustable component of the ice maker.

NOTE: Later production had the N.O. terminal removed from the water valve switch making this a single-pole single-throw switch; however, for service the same double-throw switch as used for the holding switch and shut-off switch can still be used for the water valve switch as before by not using the N.O. terminal.

SHUT-OFF SWITCH—stops operation when the storage bin is full, *Figure WW-7.*

TIMING CAM AND COUPLER

Three separate cams are combined in one molded delrin part. One end is attached to a large timing gear. The other end is coupled to the ejector. The function of the separate cams is as follows:

The inner cam operates the shut-off switch lever arm.

The center cam operates the holding switch.

The outer cam operates the water valve switch.

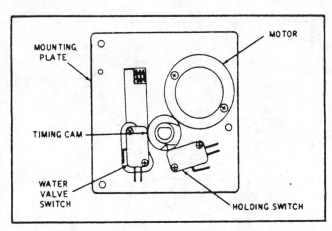

Figure WW-8, Holding Switch and Water Valve Switch

TIMING GEAR, Figure WW-9

This large molded plastic gear is driven by the motor and in turn rotates the cams and ejector. A "D" shaped mounting hole attaches to the timing cam. Spacer tabs on the backside prevent the gear from binding on the mounting plate.

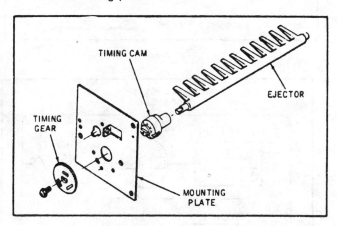

Figure WW-9, Timing Gear and Cam

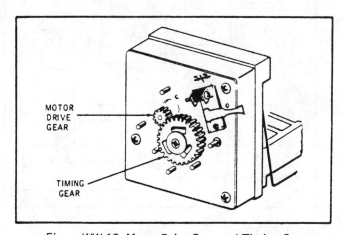

Figure WW-10, Motor Drive Gear and Timing Gears

MOTOR, Figures WW-10, 11

A low-wattage stall-type motor is used. It has a single-output shaft which drives the timing gear. This gear turns the timing cam and ejector blades at about one revolution per three minutes (1/3 RPM).

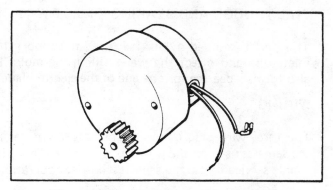

Figure WW-11, Stall Type Motor and Drive Gear

WATER VALVE, Figure WW-12

The water valve is solenoid operated and, when open, releases water from the source to the ice maker mold. The amount of water is directly proportional to the length of time the water valve switch is held closed by its timing cam.

Inside the valve is a flow washer which acts as a water pressure regulator. The water valve also incorporates a 60-mesh screen water strainer.

The solenoid coil is rated at 10 to 15 watts when energized. The coil circuit is in series with the mold heater and as a result the voltage drops to about 105 VAC at the coil.

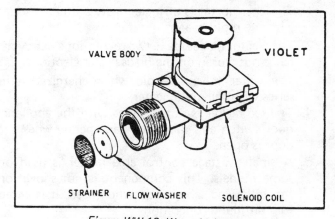

Figure WW-12, Water Valve

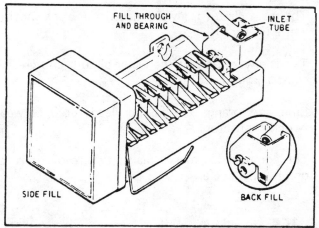

Figure WW-13, Fill Trough and Bearing

FILL TROUGH AND BEARING, Figure WW-13

This part is molded nylon. It is used to support the inlet tube and direct the water filling the mold. It also forms a bearing for one end of the ejector blades.

WIRING

The three-wire plug connector on this ice maker is the same as used on the previous ice maker. This will enable service to use the same test cord for operation.

Since the ice maker is of a compact design, the wiring inside the support housing must also be compact. Always replace wiring exactly as removed.

CYCLE OF OPERATION
(Figures WW-14 thru WW-24, or Figures WW-25 thru WW-42)

The ice maker's operation is not complex. However, an understanding of its cycle is necessary in order for a serviceman to be able to diagnose difficulty.

The following series of electrical schematics illustrates a typical cycle. Also shown are various components and their relative position during the cycle.

Note the following differences from previous ice makers:

1. The thermostat is a single-throw switch in series with the mold heater.
2. The ejector blades make two revolutions per cycle. Ice is not stored on the blades after ejection.
3. The water valve solenoid when energized is in series with the mold heater.
4. The ice maker is not in series with the appliance door switch and therefore, can operate when the door is open.
5. A small wattage control area heater is used on some models. This component is not shown on the following schematics. When used, it is wired for continuous operation.

The main circuit is always made from black (hot) to white (Neutral).

Tan is always part of the water circuit, and green is chassis ground.

Blue and orange are jumper leads used between switch connectors.

The compact ice maker makes two revolutions for one normal cycle.

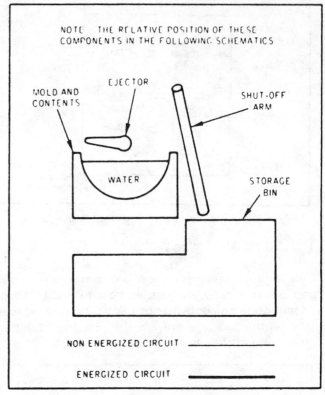

Figure WW-14

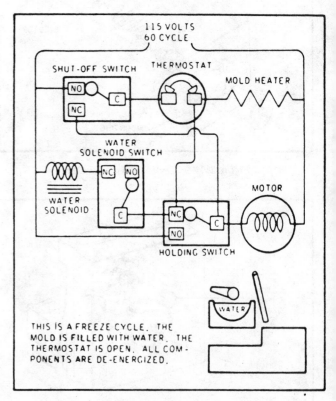

Figure WW-15

A manual test cycle makes only one revolution because the thermostat is not closed.

N.O. = Normally open
N.C. = Normally close
  C. = Common
C.C. = Cold closed 18ºF = 6º
W.C. = Warm closed 48º = 6º

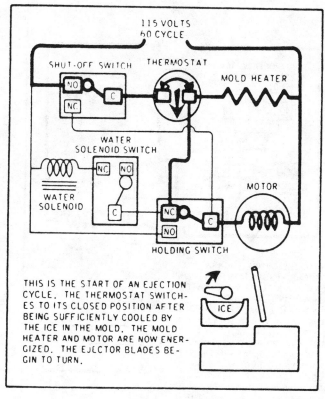

THIS IS THE START OF AN EJECTION CYCLE. THE THERMOSTAT SWITCHES TO ITS CLOSED POSITION AFTER BEING SUFFICIENTLY COOLED BY THE ICE IN THE MOLD. THE MOLD HEATER AND MOTOR ARE NOW ENERGIZED. THE EJECTOR BLADES BEGIN TO TURN.

Figure WW-16

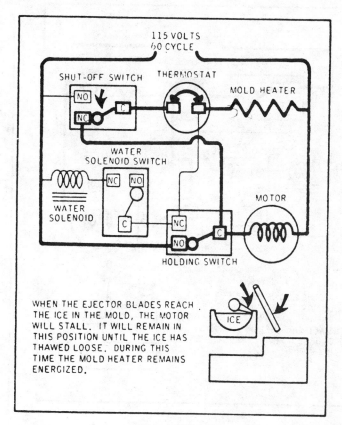

WHEN THE EJECTOR BLADES REACH THE ICE IN THE MOLD, THE MOTOR WILL STALL. IT WILL REMAIN IN THIS POSITION UNTIL THE ICE HAS THAWED LOOSE. DURING THIS TIME THE MOLD HEATER REMAINS ENERGIZED.

Figure WW-18

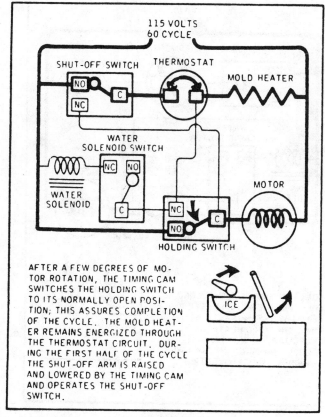

AFTER A FEW DEGREES OF MOTOR ROTATION, THE TIMING CAM SWITCHES THE HOLDING SWITCH TO ITS NORMALLY OPEN POSITION; THIS ASSURES COMPLETION OF THE CYCLE. THE MOLD HEATER REMAINS ENERGIZED THROUGH THE THERMOSTAT CIRCUIT. DURING THE FIRST HALF OF THE CYCLE THE SHUT-OFF ARM IS RAISED AND LOWERED BY THE TIMING CAM AND OPERATES THE SHUT-OFF SWITCH.

Figure WW-17

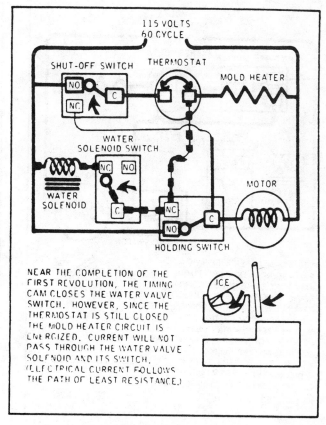

NEAR THE COMPLETION OF THE FIRST REVOLUTION, THE TIMING CAM CLOSES THE WATER VALVE SWITCH. HOWEVER, SINCE THE THERMOSTAT IS STILL CLOSED THE MOLD HEATER CIRCUIT IS ENERGIZED. CURRENT WILL NOT PASS THROUGH THE WATER VALVE SOLENOID AND ITS SWITCH. (ELECTRICAL CURRENT FOLLOWS THE PATH OF LEAST RESISTANCE.)

Figure WW-19

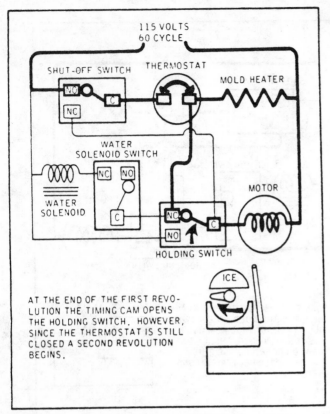

AT THE END OF THE FIRST REVO-
LUTION THE TIMING CAM OPENS
THE HOLDING SWITCH. HOWEVER,
SINCE THE THERMOSTAT IS STILL
CLOSED A SECOND REVOLUTION
BEGINS.

Figure WW-20

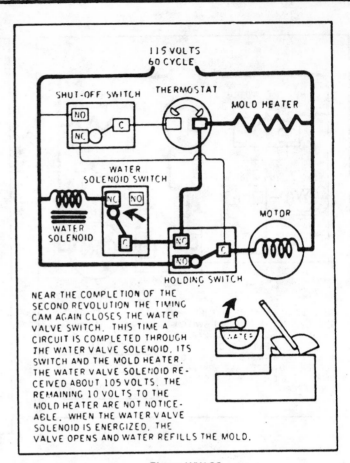

NEAR THE COMPLETION OF THE
SECOND REVOLUTION THE TIMING
CAM AGAIN CLOSES THE WATER
VALVE SWITCH. THIS TIME A
CIRCUIT IS COMPLETED THROUGH
THE WATER VALVE SOLENOID, ITS
SWITCH AND THE MOLD HEATER.
THE WATER VALVE SOLENOID RE-
CEIVED ABOUT 105 VOLTS. THE
REMAINING 10 VOLTS TO THE
MOLD HEATER ARE NOT NOTICE-
ABLE. WHEN THE WATER VALVE
SOLENOID IS ENERGIZED, THE
VALVE OPENS AND WATER REFILLS THE MOLD.

Figure WW-22

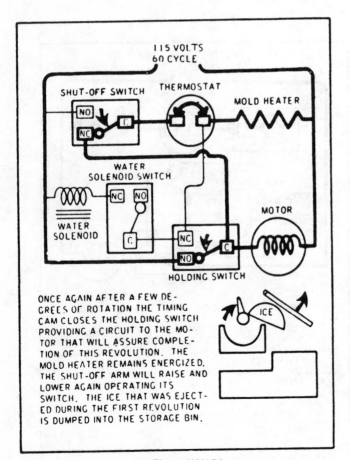

ONCE AGAIN AFTER A FEW DE-
GREES OF ROTATION THE TIMING
CAM CLOSES THE HOLDING SWITCH
PROVIDING A CIRCUIT TO THE MO-
TOR THAT WILL ASSURE COMPLE-
TION OF THIS REVOLUTION. THE
MOLD HEATER REMAINS ENERGIZED.
THE SHUT-OFF ARM WILL RAISE AND
LOWER AGAIN OPERATING ITS
SWITCH. THE ICE THAT WAS EJECT-
ED DURING THE FIRST REVOLUTION
IS DUMPED INTO THE STORAGE BIN.

Figure WW-21

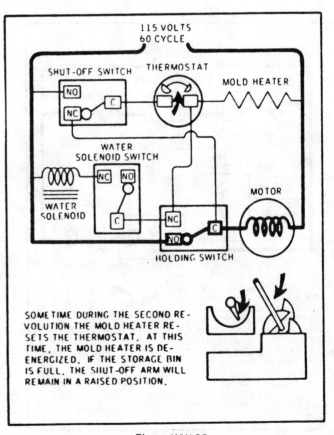

SOME TIME DURING THE SECOND RE-
VOLUTION THE MOLD HEATER RE-
SETS THE THERMOSTAT. AT THIS
TIME, THE MOLD HEATER IS DE-
ENERGIZED. IF THE STORAGE BIN
IS FULL, THE SHUT-OFF ARM WILL
REMAIN IN A RAISED POSITION.

Figure WW-23

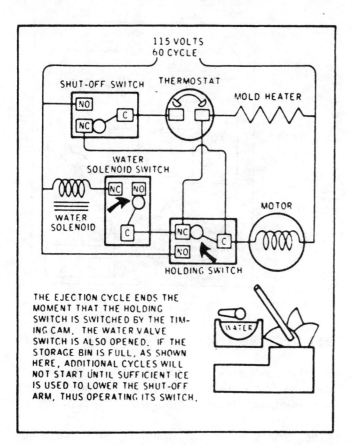

THE EJECTION CYCLE ENDS THE MOMENT THAT THE HOLDING SWITCH IS SWITCHED BY THE TIMING CAM. THE WATER VALVE SWITCH IS ALSO OPENED. IF THE STORAGE BIN IS FULL, AS SHOWN HERE, ADDITIONAL CYCLES WILL NOT START UNTIL SUFFICIENT ICE IS USED TO LOWER THE SHUT-OFF ARM, THUS OPERATING ITS SWITCH.

Figure WW-24

NOTE THE RELATIVE POSITION OF THESE COMPONENTS IN THE FOLLOWING SCHEMATICS

NON ENERGIZED CIRCUIT ————

ENERGIZED CIRCUIT ━━━

Figure WW-26

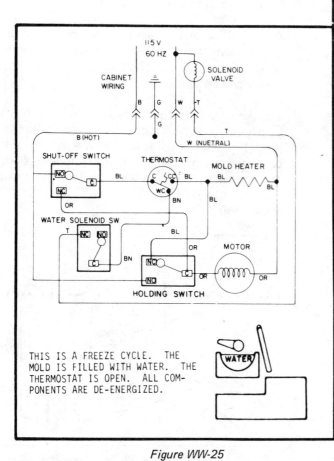

THIS IS A FREEZE CYCLE. THE MOLD IS FILLED WITH WATER. THE THERMOSTAT IS OPEN. ALL COMPONENTS ARE DE-ENERGIZED.

Figure WW-25

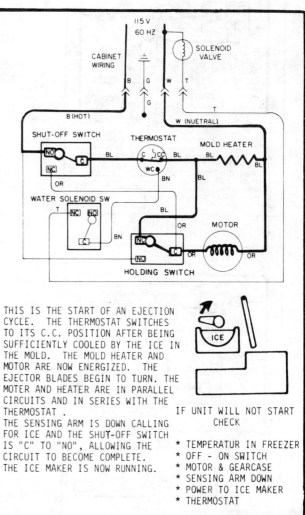

THIS IS THE START OF AN EJECTION CYCLE. THE THERMOSTAT SWITCHES TO ITS C.C. POSITION AFTER BEING SUFFICIENTLY COOLED BY THE ICE IN THE MOLD. THE MOLD HEATER AND MOTOR ARE NOW ENERGIZED. THE EJECTOR BLADES BEGIN TO TURN. THE MOTER AND HEATER ARE IN PARALLEL CIRCUITS AND IN SERIES WITH THE THERMOSTAT.
THE SENSING ARM IS DOWN CALLING FOR ICE AND THE SHUT-OFF SWITCH IS "C" TO "NO", ALLOWING THE CIRCUIT TO BECOME COMPLETE. THE ICE MAKER IS NOW RUNNING.

IF UNIT WILL NOT START CHECK

* TEMPERATUR IN FREEZER
* OFF - ON SWITCH
* MOTOR & GEARCASE
* SENSING ARM DOWN
* POWER TO ICE MAKER
* THERMOSTAT

Figure WW-27

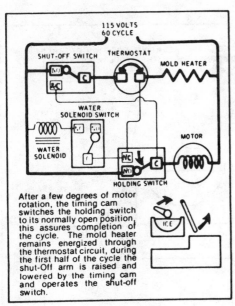

After a few degrees of motor rotation, the timing cam switches the holding switch to its normally open position, this assures completion of the cycle. The mold heater remains energized through the thermostat circuit, during the first half of the cycle the shut-off arm is raised and lowered by the timing cam and operates the shut-off switch.

Figure WW-28

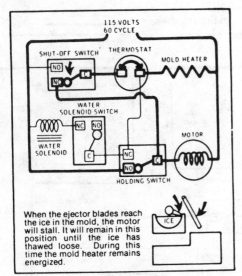

When the ejector blades reach the ice in the mold, the motor will stall. It will remain in this position until the ice has thawed loose. During this time the mold heater remains energized.

Figure WW-29

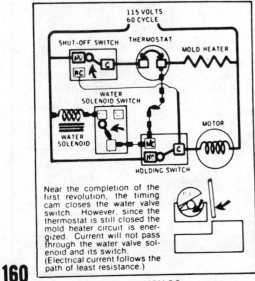

Near the completion of the first revolution, the timing cam closes the water valve switch. However, since the thermostat is still closed the mold heater circuit is energized. Current will not pass through the water valve solenoid and its switch. (Electrical current follows the path of least resistance.)

Figure WW-30

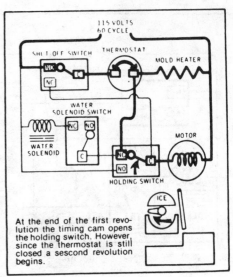

At the end of the first revolution the timing cam opens the holding switch. However, since the thermostat is still closed a second revolution begins.

Figure WW-31

TEST-CYCLING THE ICE MAKER, Figure WW-37

It may be necessary on occasion to test-cycle an ice maker to check its operation. This can be done on the repair bench or while mounted in the appliance.

If the ice maker is in an operating appliance, take precaution against condensate by allowing the cold metal components to warm before removing the front cover. The appliance cabinet wiring can be used to operate the ice maker.

A special ice maker test cord can be made for bench repair using the cabinet ice maker connector. The same cord made for the "Remote Valve" type ice maker can be used.

To manually start a cycle, first remove the front cover by prying loose with a coin at bottom of the support. Place the blade of a flat-screwdriver in the slot located on the drive gear. Turn counterclockwise until the holding switch clicks in and circuit is made to the ice maker motor. The ice maker will then complete its cycle itself.

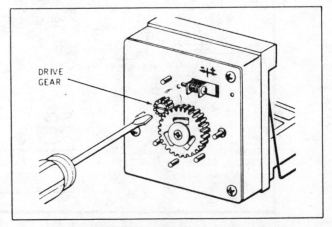

Figure WW-37, Test Cycle Manually

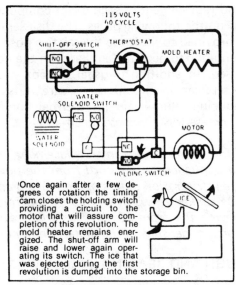

Once again after a few degrees of rotation the timing cam closes the holding switch providing a circuit to the motor that will assure completion of this revolution. The mold heater remains energized. The shut-off arm will raise and lower again operating its switch. The ice that was ejected during the first revolution is dumped into the storage bin.

Figure WW-32

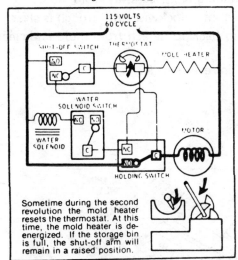

Sometime during the second revolution the mold heater resets the thermostat. At this time, the mold heater is de-energized. If the storage bin is full, the shut-off arm will remain in a raised position.

Figure WW-33

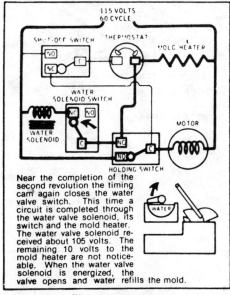

Near the completion of the second revolution the timing cam again closes the water valve switch. This time a circuit is completed through the water valve solenoid, its switch and the mold heater. The water valve solenoid received about 105 volts. The remaining 10 volts to the mold heater are not noticeable. When the water valve solenoid is energized, the valve opens and water refills the mold.

Figure WW-34

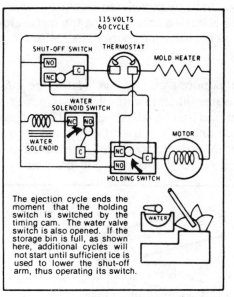

The ejection cycle ends the moment that the holding switch is switched by the timing cam. The water valve switch is also opened. If the storage bin is full, as shown here, additional cycles will not start until sufficient ice is used to lower the shut-off arm, thus operating its switch.

Figure WW-35

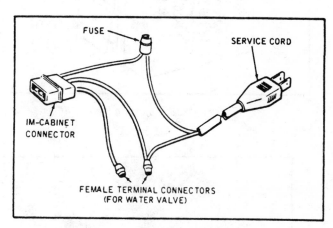

Figure WW-36, Test Cord

ADJUSTING WATER FILL, Figure WW-38

For correct water fill see Specifications. This adjustment is made at the factory on new product and need not be checked in the field except in case of very high or very low water pressure or in case of repairs involving the water valve or its switch adjustment.

To check the water fill, place a cup or other container below the inlet tube or water valve outlet. Manually cycle the ice maker and collect the water. (If air is in the water line, throw away the first few samples.)

Measure the water sample from a normal cycle in a medicine bottle or baby bottle or any vessel graduated in cc's. If the volume is more or less than specified, adjust the valve switch as required.

To adjust the valve switch, first determine how much more or less water is needed. The adjusting screw is calibrated so that one completed revolution changes the water fill about 18 cc's. Turning the screw clockwise decreases the fill while counterclockwise increases the fill.

EXAMPLE: Suppose we test cycled the ice maker and took a water fill sample of 143 cc's. We subtract 130 cc's and find that an adjustment of 13 cc's less is needed. Since one turn of the adjusting screw equals 18 cc's, then ¾ turn clockwise would reduce our fill about 13 cc's. No further checking is required.

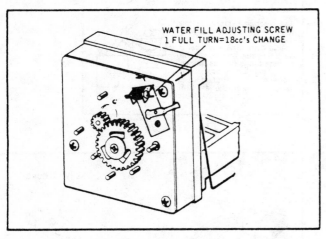

Figure WW-38, Adjusting Water Fill

PARTS REMOVAL AND REPLACEMENT

Removal and replacement of each major component is described on the following pages. The disassembly diagram is provided to illustrate the relative position of components and to become familiar with the nomenclature used. In addition to usual toolbox equipment, the serviceman should carry wire nuts for electrical splicing, alumilastic for thermal bonding and silicone grease for use where specified.

NOTE: Before attempting any replacement, disconnect the appliance service cord from the power supply. If the refrigeration unit is operating and cold, allow the ice maker to warm to room temperature before removing the front cover. This will prevent moisture from condensing on the metal components.

FILL TROUGH AND BEARING, Figure WW-13

1. Push retaining tab back away from mold.
2. Rotate counterclockwise until trough is clear.
3. Pull from back to detach from mold and ejector blades.
4. Replace in reverse order.

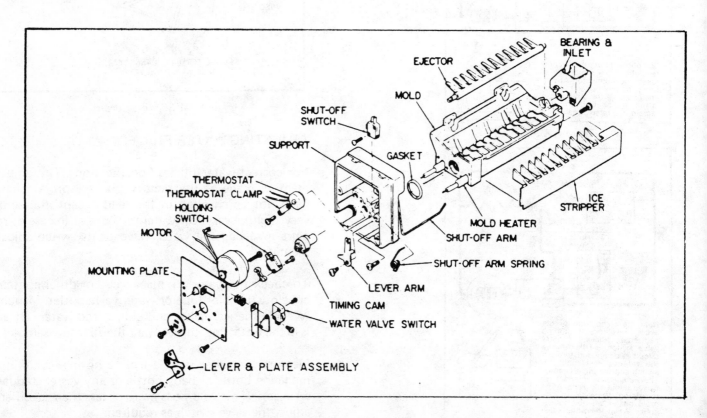

Figure WW-39, Parts Breakdown

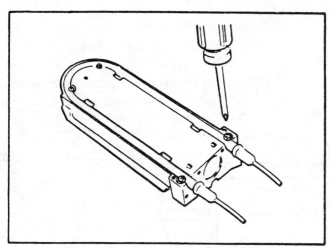

Figure WW-40, Mold Heater

CONTROL AREA HEATER

1. Remove front cover.
2. Remove mounting plate (3 screws).
3. Remove lever arm (1 screw).
4. Remove shut-off switch (2 screws).
5. Disconnect wiring and remove heater.
6. Replace in reverse order.

MOLD HEATER, Figure WW-40

1. Remove stripper (1 screw).
2. Remove front cover.
3. Remove mounting plate (3 screws).
4. Detach thermostat from mold.
5. Detach heater leads.
6. Remove mold from support (4 screws). Be careful not to destroy thermostat gasket.
7. With a flat-bladed screwdriver, pry defective heater from bottom of mold.
8. Clean old alumilastic from groove in bottom of mold.
9. Apply new alumilastic to mold.
10. Install replacement heater retaining with 4 screws (provided in service heater kit) in holes adjacent to heater.
11. Replace parts in reverse order of removal.

Be sure thermostat gasket is in place. Bond thermostat to mold with new alumilastic.

EJECTOR BLADES, Figure WW-4

1. Remove fill trough and bearing.
2. Force back and up to detach from front bearing.
3. Place small amount of silicone grease on bearing ends of replacement.
4. Replace in reverse order noting that blades are in same position as original.

ICE STRIPPER, Figure WW-3

1. Remove ice maker from cabinet.
2. Remove retaining screw at back of mold.
3. Pull stripper back to disengage from front of mold.
4. Replace in reverse order.

FRONT COVER

1. Be sure ice maker mechansim is not cold.
2. Place coin in slot at bottom of mold support and pry cover loose.
3. To replace, be sure retaining tabs inside cover are located on top and bottom, then snap in place.

MOUNTING PLATE

1. Remove front cover.
2. Remove 3 retaining screws holding plate in place.
3. Carefully remove plate disengaging end of shut-off arm and noting relative position of shut-off arm spring.
4. Before replacing plate be sure all wiring is orderly and shut-off arm spring is in place.
5. Replace in reverse order.

MOTOR, Figure WW-11

1. Remove front cover.
2. Remove mounting plate (3 screws).
3. Disconnect wiring.
4. Remove motor (2 screws).
5. Replace in reverse order.

WATER VALVE SWITCH, Refer to Figure WW-42

1. Remove front cover.
2. Remove mounting plate (3 screws).
3. Disconnect wiring.
4. Remove switch (2 screws).
5. Replace in reverse order making sure switch insulator is in place.
6. Check water fill and adjust if required.

HOLDING SWITCH, Refer to Figure WW-42

1. Remove front cover.
2. Remove mounting plate (3 screws).
3. Disconnect wiring.
4. Remove switch (2 screws).
5. Replace in reverse order making sure switch insulation is in place.

SHUT-OFF SWITCH, Refer to Figure WW-42

1. Remove front cover.
2. Remove mounting plate (3 screws).
3. Raise shut-off arm.
4. Disconnect wiring.
5. Remove switch (2 screws).
6. Replace in reverse order.

THERMOSTAT, Refer to Figure WW-42

1. Remove front cover.
2. Remove mounting plate (3 screws).
3. Loosen thermostat clip mounting screws.
4. Disconnect wiring and remove thermostat.
5. Apply alumilastic to sensing surface of replacement thermostat and bond to mold.
6. Replace in reverse order.

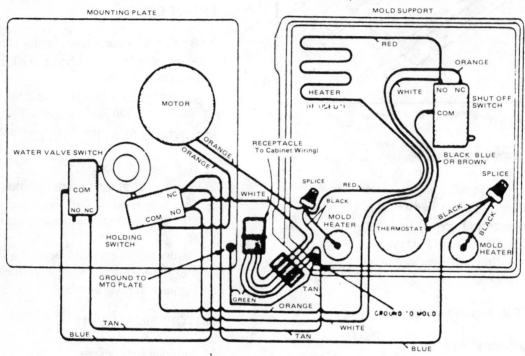

Figure WW-41, Wiring Diagram

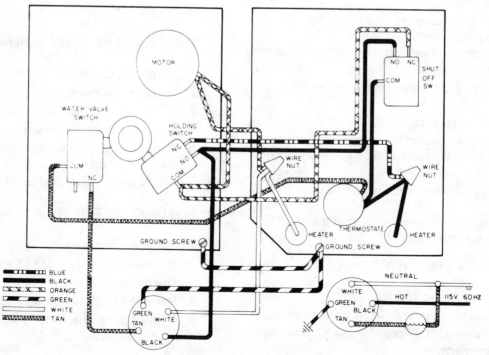

115 V. 60 HZ

Figure WW-42, Wiring Diagram

TROUBLE DIAGNOSIS AND CHECKING PROCEDURE

COMPLAINT	CAUSE	CORRECTION
Automatic Icemaker Fails To Start	Shut-off arm.	Check that arm is in lowest position. Lower arm if raised.
	No electrical power.	Check for power at black and white ice maker leads. Correct cabinet wiring if defective.
	Not cold enough.	Check mold temperature at a mounting screw. If above 10°, evaporator is not cold enough.
	Thermostat.	If mold is below 10°, manually start ice maker by pushing timing gear. If motor starts, thermostat, shut-off switch or holding switch is defective. C
	Holding switch.	With ejector blades in starting position, check terminals "C" and "NC" for continuity. Replace switch if open.
	Shut-off arm. Switch	Check that linkage is proper, adjust if necessary. Check terminals "NO" and "C" for continuity with arm in lowest position. Replace switch if open.
	Motor	Check operation with test cord. Replace motor if it fails to start.
Automatic Icemaker Fails To Complete Cycle	Holding Switch. If blades are 10 o'clock position.	With switch plunger depressed, check terminals "C" and "NO" continuity. Replace switch if open.
	Shut-off switch. If blades are in 12 o'clock position.	Check terminals "C" and "NC" for continuity. Replace switch if open.
	Mold heater or thermostat. If blades are in 4 o'clock position.	Check heater for continuity. Replace heater if open. If heater shows continuity replace thermostat.
	Motor.	Check operation with test cord. Replace motor if it fails to start.
Automatic Icemaker Fails To Stop At End of Cycle	Holding switch.	With ejector blades in starting position, check terminals "C" and "NO" for continuity. Replace switch if closed.
Automatic Icemaker Continues To Eject When Bin Is Full	Shut-off arm.	Check that linkage is proper. Switch should open when arm is in raised position. Adjust if required. Check terminals "C" and "NO" for continuity with arm raised. Replace switch if closed.

TROUBLE DIAGNOSIS AND CHECKING PROCEDURE

COMPLAINT	CAUSE	CORRECTION
Automatic Icemaker Produces Undersize Ice Pieces	Mold.	Check for level. Adjust if required.
	Water supply.	Check that supply line and water valve strainer are completely open and that adequate pressure is maintained. Clear restrictions or advise customers accordingly.
	Water valve switch.	Test-cycle IM and measure water fill. Adjust switch if required.
	Thermostat short-cycling indicated by ice shells or hollow ice in storage bin.	Check thermostat bond to mold. Assure good thermal contact with alumilastic. Check thermostat calibration by replacing with new part.
Automatic Icemaker Spills Water From Mold	Mold.	Check for level. Adjust if required. Check top edge for evidence of siphoning. Prevent capillary action using silicone grease in this area.
	Water inlet tube.	Check that inlet tube and fill trough fit properly and water does not leak during fill cycle. Adjust fit if required.
	Water valve.	Check that water does ont enter mold after cycle is completed. Replace valve if leaking and water pressure is proper.
	Water valve switch.	Test-cycle IM and check that water fill does not exceed volume capacity of mold. Adjust switch if required. With ejector blades in starting position, check terminals "C" and "NC". Replace switch if closed.
	IM fails to stop at end of cycle.	Refer to complaint "Fails to stop at end of cycle".
	Thermostat short-cycling.	Refer to complaint "undersize ice pieces."
Water Fails To Enter Mold During Cycle	Water supply.	Check that water line, valve, and valve strainer are open. Remove restrictions, open valve, or instruct customers accordingly.
	Water valve.	Observe inlet tube and fill trough for ice. If obstructed with ice; check water valve for slow leak. If valve leaks, check for proper water line pressure. Replace heater if open. Replace water valve if pressure is within specifications.
	Valve solenoid coil.	Check terminals for continuity. Replace coil if open or shorted.
	Water valve switch.	With plunger out check terminals "C" and "NC" for continuity. Replace switch if open.

NOTES

GEM PRODUCTS, INC. MASTER PUBLICATIONS

REPAIR-MASTER® For Automatic Washers, Dryers & Dishwashers

These Repair-Masters® offer a quick and handy reference for the diagnosis and correction of service problems encountered on home appliances. They contain many representative illustrations, diagrams and photographs to clearly show the various components and their service procedure.

Diagnosis and repair charts provide step-by-step detailed procedures and instruction to solve the most intricate problems encountered in the repair of washers, dryers and dishwashers.

These problems range from timer calibrations to complete transmission repair, all of which are defined and explained in easy to read terms.

The Repair-Masters® are continually updated to note the latest changes or modifications in design or original parts. These changes and modifications are explained in their service context to keep the serviceman abreast of the latest developments in the industry.

AUTOMATIC WASHERS
9001 Whirlpool
9003 General Electric
9005 Westinghouse Front Loading
9009 Frigidaire
9010 Maytag
9011 Philco-Bendix Front Loading
9012 Speed Queen
9015 Norge – Plus Capacity
9016 Frigidaire Roller-Matic
9017 Westinghouse Top Loading

DISHWASHERS
5552 General Electric
5553 Kitchenaid
5554 Westinghouse
5555 D & M
5556 Whirlpool
5557 Frigidare

CLOTHES DRYERS
8051 Whirlpool-Kenmore
8052 General Electric
8053 Hamilton
8055 Maytag
8056 Westinghouse
8057 Speed Queen
8058 Franklin
8059 Frigidaire

*The D & M Dishwasher Repair-Masters covers the following brands: Admiral — Caloric — Chambers — Frigidaire (Portable) — Gaffers and Sattler — Gibson — Kelvinator — Kenmore — Magic Chef — Magic Maid — Norge — Philco — Pioneer — Preway — Roper — Wedgewood — Westinghouse (Portable)

The Repair-Masters® series will take the guesswork out of service repairs.

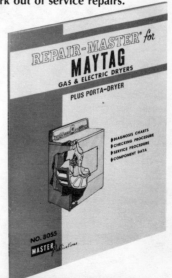

REPAIR-MASTER® for Domestic Refrigeration

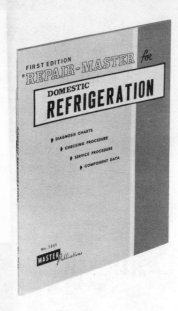

A BASIC GUIDE FOR REFRIGERATION REPAIR.

Takes the guesswork out of all your refrigeration repair work. The 7551 Repair-Master® covers the electrical, mechanical and hermetic systems of refrigeration. With the use of the diagnosis charts, malfunctions can be located and repaired rapidly.

- CHECKING PROCEDURE
- SERVICE PROCEDURE
- DIAGNOSIS CHARTS
- COMPONENT DATA

Valuable Information for the Refrigeration Servicetech!

**7551
132 Pages**

REPAIR-MASTER® for Domestic/Commercial Refrigeration

A most comprehensive Repair-Master® for the diagnosis and repair of both domestic and commercial refrigeration.

Includes electric testing, brazing, evacuation and recharging the refrigeration system. Step-by-step procedures for replacing components. Covers capillary tubes and expansion valves. Includes principles, cycle and the history of mechanical refrigeration.

This Repair-Master® is written in a manner that can be easily understood by the beginner and also serves as a complete refresher for the journeyman. A most comprehensive Repair-Master® that can be used for teaching refrigeration service.

**7552
203 Pages**

- DIAGNOSIS CHARTS
- USE OF REFRIGERATION TOOLS
- HISTORY OF MECHANICAL REFRIGERATION
- PRINCIPLES OF REFRIGERATION
- THE REFRIGERATION CYCLE

- REFRIGERATION BY HEAT
- THE ABSORPTION SYSTEM
- THE GAS REFRIGERATOR
- COMPONENT REPLACEMENT
  AND MUCH MORE IS INCLUDED IN THIS REPAIR MASTER

REPAIR-MASTER® For Window Air Conditioners

This Repair-Master® offers a quick and handy reference for the diagnosis and correction of service problems. Many illustrations, diagrams and photographs are included to clearly show the various components and their service procedures.

A Complete Service and Repair Guide!

PRINCIPLES OF REFRIGERATION
- DIAGNOSIS CHARTS
- TEST CORD TESTING
- PROCEDURES
- STEP-BY-STEP REPAIR
- SHOP PROCEDURES

7541
84 Pages

REPAIR MASTER® For Domestic Automatic Icemakers

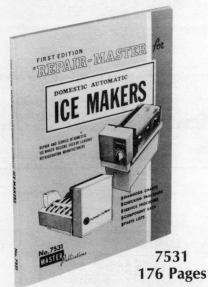

Includes ten different designs of Automatic Ice Makers presently found in domestic refrigerators.

A special section is included for Whirlpool design self-contained Ice Maker. Illustrations, charts and new parts design information makes this Repair-Master® an important tool.

- DIAGNOSIS CHARTS
- TESTING PROCEDURES
- ADJUSTMENT PROCEDURES
- STEP-BY-STEP REPAIRS

7531
176 Pages

TECH-MASTER® For Refrigerators & Freeze

For All Major Brands

Offering a quick reference by brand, year, and model number.

Handy 8½" x 5½" book is designed to be easily carried in the tool box or service truck. The first time you use these books they will more than repay the nominal purchase price.

Operating Information for Over 15,000 Models
- TEMPERATURE CONTROL PART NUMBERS
- CUT-IN AND CUT-OUT SETTINGS
- UNIT HORSEPOWER
- REFRIGERANT AND OIL CHARGE IN OUNCES
- SUCTION AND HEAD PRESSURES
- START AND RUN WATTAGES
- COMPRESSOR TERMINAL HOOK-UPS
- DEFROST HEATER WATTAGES

No. 7550	No. 7550-2	No. 7550-3	No. 7550-
1950-1965	1966-1970	1971-1975	1976-1980
429 Pages	377 Pages	332 Pages	378 Pages